Achieving Success with the Engineering Dissertation

Petra Gratton · Guy Gratton

Achieving Success with the Engineering Dissertation

 Springer

Petra Gratton
Department of Mechanical
and Aerospace Engineering
Brunel University London
Uxbridge, Middlesex, UK

Guy Gratton
School of Aerospace
Transport and Manufacturing
Cranfield University
Cranfield, Bedfordshire, UK

ISBN 978-3-030-33194-8   ISBN 978-3-030-33192-4   (eBook)
https://doi.org/10.1007/978-3-030-33192-4

This Springer imprint is published by the registered company Springer Nature Switzerland AG
The registered company address is: Gewerbestrasse 11, 6330 Cham, Switzerland

Preface

The Dissertation and This Book

The high point of any engineering degree is the dissertation, also called variously the "major project", "capstone project" and a few other things. This is a step forward in working as an engineering professional: being given a task to solve, where there's not a clear method to complete it, nor a known right answer. This can be both exciting and frightening; engineers train to be able to solve real engineering problems, but actually doing so, not knowing what the answer is, isn't easy.

We are both engineers, and years ago undertook dissertations on our own degrees at Brunel University, the University of Southampton and the University of Surrey. After entering industry, somehow, we both got lured back into academia, which has resulted in supervising and marking many hundreds of dissertations over nearly four decades between us. This book is the one we wanted for our own students, offering the best advice we can give on not just how to do well in the dissertation, but how to use it as a springboard into a successful career.

In preparing this, we'd like to thank our families and our many dissertation students—we've learned about the topic ourselves from every one of them.

The Concepts in *Achieving Success*

Everybody wants dissertations to go well, and they are also a key part of developing as a professional engineer. That involves preparing before the project, and using what was done and learned into the job search and whilst starting the first job.

Because engineering projects are team efforts, we have focused on the important relationship between the student and their supervisor—and also, *the project client*—somebody who is providing support and has an active interest in the project outcome. There won't always be a client, or sometimes either the student or the supervisor will "wear two hats" and take that role, but it's a very useful concept.

There are many engineering disciplines, and we've tried to vary from which disciplines examples are drawn; if these aren't in your own field—please read and try and learn from them anyhow: good practices are usually universal.

The Illustrations

Engineering writing needs illustrations, equations, graphs, drawings and pictures. *Achieving Success* uses our own diagrams and photographs, as well as a few from other sources. We wanted to make this book visually attractive and enjoyable to read; with the help of the Heath Robinson Museum in Pinner, we've used a lot of illustrations from William Heath Robinson, whose name is synonymous with complex engineering and knotted string, and whose brilliant art is well known to many engineers. Clearly, pictures like Fig. 1 aren't here as serious illustrations of what you should be doing in your university work, but they are there to improve the readability, give you something to think about, and we hope you enjoy them. If you'd like to learn a little more about Heath Robinson's life and work, turn to Appendix D. (If it's not obvious—we're fans!) If you'd like to learn more, there are also many other books of his work, as well as the museum, and it's not unusual to see exhibitions of his art appear around the world.

How to Use *Achieving Success*

Ideally, start using this book the year before starting the dissertation—Chaps. 1 to 4 cover preparation for the project. Appendix C also will show how a project can be used to enter national or international competitions. Chapters 5 to 8 are for the early part of the project, once there's a topic and supervisor. Students might like to suggest that their supervisor and (if there is one) client read the table of contents and Chaps. 4–8, although these cover more ground than any single project, so will need using selectively.

Chapters 9 to 14 cover the central core of the project work and report, towards the final report presentation. There's a lot of detail that here users should expect to jump around these sections—as well as, for example, probably using Appendix A on referencing.

Finally, Chaps. 15 and 16 are about how to use the dissertation to improve your chances of entering a good graduate career, and then about the possibility of publishing your work outside the University course.

Appendices A–C are there for reference throughout the project—and other parts of any engineering degree course. Appendix D is about William Heath Robinson.

Achieving Success is written in a fairly informal style—this is to make it more readable, but of course a dissertation report itself needs to be much more formal.

Fig. 1 Mr William Heath Robinson's Welsh rarebit (cheese-on-toast) machine: fuel for the dissertation?

Good Fortune

We wish you success with your dissertation, your degree and your onwards career. We've tried to give you the best advice we can in this book, and hope it's good. We would love to hear your feedback about how useful it was, how you've used it, and anything you think could be improved. We're very easy to contact through either Brunel University London, Cranfield University or the publisher. We also tweet about the subject matter at @ASEDtweets.

Uxbridge, UK   Petra Gratton
Cranfield, UK   Guy Gratton
August 2019

Contents

Chapter 1
What is the Dissertation?

Abstract The dissertation engages the engineering student with professional engineering practice, and develops the skill of working under their own direction on a large project. It will involve working with uncertainty: solving problems that don't have prior known solutions. There are six main types of project: computation-based design feasibility studies, experimental, theoretical and test and evaluation There will be three people mainly involved in a dissertation project: the student (who will do most of the work), the supervisor and for many projects, a client. This chapter also defines the main terminology used in dissertation projects.

© Springer Nature Switzerland AG 2020
P. Gratton and G. Gratton, *Achieving Success with the Engineering Dissertation*,
https://doi.org/10.1007/978-3-030-33192-4_1

1.1 Purpose and Structure

1.1.1 What's a Dissertation for?

The dissertation is both an opportunity to learn and to demonstrate the knowledge, know-how and skills you have acquired so far as an engineer. The dissertation will give you chances to fulfil the following:

(1) To learn how, and then demonstrate your ability to work as a professional engineer—particularly including the ability to use multiple engineering subjects together.
(2) To show you can learn things without having someone to explicitly teach them to you.
(3) To give you experience of working on questions with no single right answer (this is sometimes called "working with uncertainty").
(4) To show you can work as part of a team (particularly, but not only for group projects).
(5) To learn how and then show that you can produce a large high-quality written engineering report.

It's important to read your university's description of the *anticipated learning outcomes* for your dissertation project. These are often contained in a document called something like a "Module Descriptor", which you may have been provided with, or can find on the university intranet. It will usually contain descriptions of what markers will be looking for eventually in your submissions.

1.1.2 Dealing with Uncertainty

Working with uncertainty is characteristic of engineering. So, the nature of dissertation projects means that they will contain an element of dealing with unknown information, or using data that is known to be incomplete or contain some uncertainty. Indeed, this is an essential lesson that every engineer needs to learn. Whatever the type of project that is chosen, assumptions must be made, and these need to be clearly stated in the report.

One of the key differences between Bachelors and Masters degree-level dissertations is the amount of uncertainty and how it is dealt with. If you are to demonstrate your mastery of a topic, you need to demonstrate initiative, ingenuity and independence. The dissertation for every degree is a training process. The thinking process is going to be the same, and the narrative arc of the report will be the same, but the treatment of the content may be subtly different—demanding more independence of thought and originality for higher degrees.

1.1.3   The Structure of the Project

Dissertation projects usually centre on the student (or if they are group projects a small group of students), working under the oversight of a supervisor—who will usually be a member of the university's academic staff. In some cases there will be a project client—a person or organization who provides input into the project because they want to see the outputs the project promises. Certainly, the student(s) will be doing most of the work and the student(s) has (have) the greatest vested interests in a good outcome.

Most commonly, project titles are offered to students at the start of the academic year, or at the end of the previous year: also likely, there will be an opportunity for students to propose their own. Students will be matched with projects and supervisors and a timeline will be laid down for the completion of the project.

Typically, a project will start with a literature review where the student investigates what is presently known about the topic, in parallel with their creating a project plan showing how they will deliver the planned outputs on time, what resources they'll need and so on. The literature review and plan will always need to be discussed and agreed with the supervisor but often there'll be a requirement to submit them for marking—if there is that will be in the university's dissertation guide documents. The literature review will certainly also be required to go later into the final report.

Then, following the plan, students will progress through their project usually with regular reviews of the plan and progress meetings—particularly between the student and supervisor. Depending upon the nature of the project time may be split between data gathering, experimental work, meeting and interviewing people with interests in the subject matter, analysis and writing the report. All of these and the relative proportions will depend upon the project.

At the end, normally the student(s) will submit their report for assessment, which will be independently marked by both their supervisor and one or more independent academics. It's also common that the student will be asked to make a short presentation during assessment, followed by a short question and answer session, about their dissertation.

You should have been told what proportion of the marks for the year and degree comes from the dissertation, and the split of the various parts of the dissertation in terms of marks. It's reasonable to assume that the total effort should be split roughly in proportion to the marks on offer—but equally much of the effort will support the whole of the project so that's not a strict rule. Some marks may be available for effort and conduct (usually as judged by the supervisor), but mostly it will be the submitted work and any presentations that count highest.

It is important to remember throughout all of this that the dissertation is both a piece of assessed work and a learning and development opportunity. Everybody should remember that what the student learns from it (and often the connections they make) is, in the long run, at least as important for the student as the marks they get. This is not just a glorified examination.

1.1.4 The Structure of the Report

This will be discussed in much more detail later in Sect. 14.1. It's likely that the university will give you a document guiding the general structure of the report and if that contradicts what we say here, use the university rules. Having said that, there is not much variation in understanding of best practice anywhere, which usually looks like as described in Table 1.1.

Your university's guidelines should have provided a limit (and possibly a target) length for the overall report. It's important to stick to these—and remember that (contrary to what many students seem to think) markers rely upon the quality and content of a report, not a ruler and wordcount, when determining what they think of it. American author Mark Twain once wrote "I apologize for such a long letter—I didn't have time to write a short one" and this should always be remembered—failing to keep a report within limits (and ideally well within, a limit is not the same as a target) is discourteous to the reader, and if the reader is also a marker, it is likely to annoy them and attract bad marks.

1.2 Types of Project

1.2.1 What Types of Projects Are Suitable for an Engineering Dissertation?

There are a variety of project-types that are suitable for an engineering dissertation. Some universities, or courses, will impose conditions upon their engineering projects, or upon specific disciplines of engineering, but here we'll try to highlight a selection of those possible types.

1. Computational-based investigation (modelling)
2. Design
3. Feasibility study
4. Physical experiment
5. Theoretical research
6. Test and evaluation.

It is possible to do a project that is some combination of the above types. The key element to the dissertation is that it contains some investigation or research. This ensures some originality. Generally speaking, the higher the degree, the more originality needs to be demonstrated in the dissertation. For sub-honours degree, you may be permitted to buy a kit (car, aircraft, robot, house, etc.) and build it; write a report about the build-experience; job done! At honours degree-level, you are likely to be expected to improve upon the design (at least), or evaluate its performance, or demonstrate how the kit could be adapted in some way. There needs to be something novel in what your project proposes. Modelling the airflow over a winglet might

Table 1.1 Description of a generic structure for a dissertation report

Abstract	This is usually the first item in the report, and the last to be written; it is (usually) a single paragraph summarizing what the report says, including a very brief summary of the conclusions and any recommendations
Acknowledgments	It's common, but not usually mandatory, to include a short acknowledgment section stating and thanking whoever has been crucial to the success of the project. Whether to include this, and what to say, is normally entirely down to the student: entirely a personal decision unlikely to affect any other outcome
Introduction	A short section introducing the topic, explaining why the dissertation was conducted and giving a short outline of the structure of what is to come in the report
The literature review	A review, with references, and discussion of the meaning and context of those references, of what was already there to be found out about the topic. This may often also include references detailing how the dissertation would be conducted—for example, guides to analysis or experimental work (or even this book!)
The body text	This is the largest part of the report likely to include a combination of data gathering (from surveys, experimental work, archival material, etc.) and analysis. It's very much representative of working as a junior professional engineer, in a sub-field relevant to the specific project. This will probably be several chapters, with each having a relevant heading such as "User survey", "Experimental work", "Deriving a theoretical model"—or more likely much more specific titles such as "Designing the powertrain", "Testing the prototype user interface" or "Carrying out the building survey". Students and supervisors are likely to have detailed discussions about that structure
Conclusions	Normally the flow of writing leads to conclusions, perhaps about how a problem can be solved, or about the characteristics of something being assessed. In some reports, this might essentiallly be a short summary of what's been determined
Recommendations	Not always in every report. It's common that leading from the conclusions, the report makes recommendations. This may be about how to improve something, incorporate the work into an ongoing activity or for some subsequent "further research"
References	The references cited in the body of the report will normally be listed together near the end of the report
Figures and Tables	Some reporting standards will still (as was once normal, particularly before word processors were in widespread use) collect all of the illustrations together at the end of the report. More likely nowadays, however, this won't be a separate section and these will all be embedded into the body text
Appendices	If there is any more information which should be added to the report—for example critical interview transcripts, computer programs or expanded laboratory results—then these can be added as discrete appendices at the end. However, any appendix should be both self-contained and non-essential to the reader's ability to understand the main report. Before deciding to include any appendices, always check whether these sit within, or additionally to, the main report length limits

provide you with the opportunity to learn to use a computational-fluid-dynamic (CFD) software package and produce some pretty multi-coloured pictures, but it does not (of itself) break new ground. So, if you take a project idea to a supervisor, be prepared to answer the question: where is the originality in this?

1.2.2 Computational-Based Investigation Projects

These projects are popular with students and academic staff alike. This popularity may be due to the belief that learning to use simulation software to model something using finite-element analysis (FEA) or computational fluid dynamics (CFD) will improve employment prospects. This may be true for some employers, but it's worth remembering that not all jobs involve computer modelling.

1.2.3 Design Projects

This umbrella term can be divided into at least four further sub-categories of projects:

(i)   *Product design*—this is a process well-understood and documented in various textbooks. It is essentially an iterative process of investigating a need (or want), proposing competing solutions, critically evaluating their merits and deciding upon a conceptual design of a product to satisfy the initial requirement. During the process a product design specification (PDS) should be drafted, and it may be that final product is visualized in a computer-generated 3-D model, or even a mock-up or full prototype created as part of the assessment.

(ii) *Vehicle design/configuration study*—this is usually carried out as a conceptual design, where characteristics, such as size, shape, power, weight, performance and cost, are established based upon design requirements and available technology.

(iii) *Design-and-build*—this might be considered as a progression of a product design or vehicle design project, for example, where a prototype of the physical artefact is required as a performative element of the project assessment, that is, a proportion of the final mark is awarded to demonstrating that the thing works as expected.

(iv) *Design, make and evaluate (DME)*—this might be considered as a further progression of the design-and-build project. It is more likely to be acceptable at Master's level, as the evaluation of performance of the artefact requires the necessary criticality. It should be noted that although these projects are usually referred to as DME, they are frequently done in the order of EDM. In these cases, the performance of the subject is critically evaluated, so that the design and manufacture of the subject is better informed of previous problems encountered. These types of projects can clearly be the basis of interuniversity competitions, such as the Formula Student or UAV Challenge (see Appendix C), as extensive testing should be carried out before introducing the object to competitive scrutiny.

1.2.4   Feasibility Studies

These types of project consider the application of existing technology to a novel situation to determine what compromises, or adjustments, need to be overcome in order to get something to work, or make something happen. This is frequently accompanied by a cost–benefit analysis, which addresses the financial implications of the thing-being-studied throughout its lifetime. In industry, consulting engineers frequently carry out feasibility studies, so this is a good type of project for developing the life-skills that will be used in an engineering career. The comparison of multiple factors necessary for evaluating feasibility lend this type of project to the use of a figure of merit (FoM) calculation, which, when used well, can be used to demonstrate a high level of understanding of the complex problem(s) being studied. Examples of feasibility studies that we have supervised have included *Selection of a Replacement Aircraft for Airborne Atmospheric Research*, and *Application of Renewable Technologies to a University Hall of Residence Building*.

1.2.5   Physical Experiments

These are probably best considered as the compliment to the computer-based investigation. Experimental design is discussed further in Sect. 7.2. A physical experiment will usually involve demonstration of understanding the underlying theory, development of an experiment, building (or adaptation) of a test-rig, running of the experiment for the collection of data and analysis of collected data against theory. Often, these types of projects can have far-ranging applications: think of a property of a thing, and then think about how that property can be characterized. For example, something as abstract as a shape can have aerodynamic properties, and something as concrete as a component in a heart-valve can have material properties such as strength. Depending upon what the characteristic is and what the thing is will depend upon how the investigation into its characteristics is carried out.

1.2.6   Theoretical Research Projects

These projects involve doing research using literature, prior data and "longhand" or computational analysis to predict something. Not unlike a computational project, this, very commonly, may be combined with some experimental work so as to both predict something and validate those predictions. "Validate" here does not mean "prove true" but is more about testing and refining the theoretical results.

1.2.7 Test and Evaluation (T&E) Projects

These projects take an existing product and assess it with one of two objectives in mind:

(i) Development T&E—this type is about how to develop a product into one that is needed or wanted by its market.
(ii) Operational T&E—this is about how to get the best out of an existing product.

Undertaking a T&E project can be quite challenging as this type of project frequently uses a variety of both quantitative and qualitative data techniques. Indeed, there is great scope for developing some useful skills much in demand by prospective employers. As T&E projects are so widely used in industry, there may be scope for collaborating with an industrial sponsor, which could result in a job offer!

1.3 The Project Client

1.3.1 Introduction

Engineering projects are all done for a reason—and for someone who owns that reason. That person is the client, and the best dissertations have a client as well. This is a person or organization who wants the outcomes that the project can provide. Here are examples of a project client:

- A competition organizer who has set down rules and deadlines for an engineering project competition.
- A disabled person who would like a device designing and making to improve their quality of life.
- An academic (most likely also the project supervisor) who wants specific answers to support an ongoing research project.
- The student themselves, who wants to develop an idea towards starting a company after they've graduated.

This, of course, parallels the role of the client in most engineering businesses (and equally, not-for-profit organizations) where the client is being provided a specific service, and that has to be done by a deadline, and within a predefined budget.

You can complicate this by adding in that most engineering projects (both academic and professional) have multiple stakeholders. Whilst all clients are stakeholders (as in, they have a clear interest in the outcomes), not all stakeholders are clients

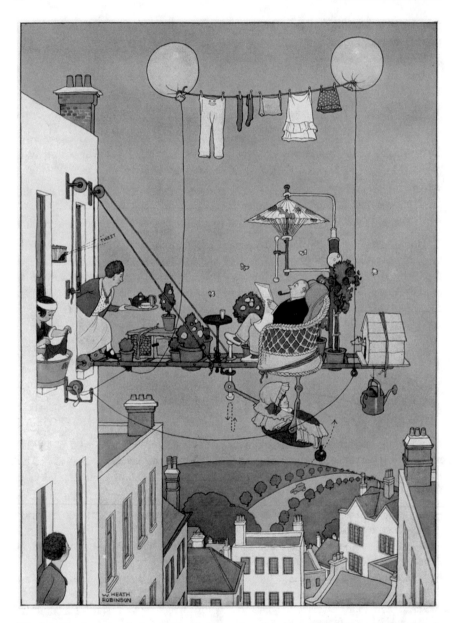

Fig. 1.1 The Deckcheyrie for unbalconied flats—possibly a result of failing to limit a client's expectations (by Mr William Heath Robinson)

(some stakeholders may not be interested in making an outcome happen—only in the consequences of it happening). For example, if designing a new apartment building, the organization paying for the building is the client—but the eventual residents, the building standards inspectors and the neighbours of the building site are all stakeholders (see the next chapter). So here, we'll separate them out.

1.3.2 Engaging the Client

The dissertation's client is the person or organization who wants the project delivered with specific outcomes. The student under the guidance of their supervisor must establish the following:-

(1) Who the client is (there may be several, but usually not)
(2) What the client wants from the project
(3) That the client's desires can be met within the timescale and resources of the dissertation—and if this can't be done, then that objectives have been changed until they are realistic (Fig. 1.1)
(4) The mechanism by which the student can communicate with the project client about what is being delivered and how.

It's important to realize that in most cases the client is not there to dictate how the project will be run. That is for the student, their supervisor and the university—to coin an old phrase "if you have bought a dog, stop barking yourself". If the student or their supervisor is the client, however, they may need to compartmentalize their multiple roles and be clear when they are acting as client, and when they are acting as student or supervisor. Trying to take two roles at the same time can cause confusion, and potentially a confused and poor outcome. There are various ways to prevent this, but mostly it will probably need to be based upon early and clear definition of the project aims and objectives by the "internal client", who then steps sideways into their role as student or supervisor solely, for as long as the project aims remain apparently achievable.

1.4 Whose Project is it?

So, the dissertation project and the eventual report. Whose is it? Who really cares about how well it goes, and the outcomes?

Well, the answer is everybody owns it. The student of course wants to produce a good piece of work that'll be a credit to them and help carry them into the start of their career as a professional engineer or researcher—and if they have the right attitude, they'll want to learn as much from it as possible too.

The supervisor should always want their student to do well, and to see them learn as much as possible: but a potential conflict here is that they're also not allowed to

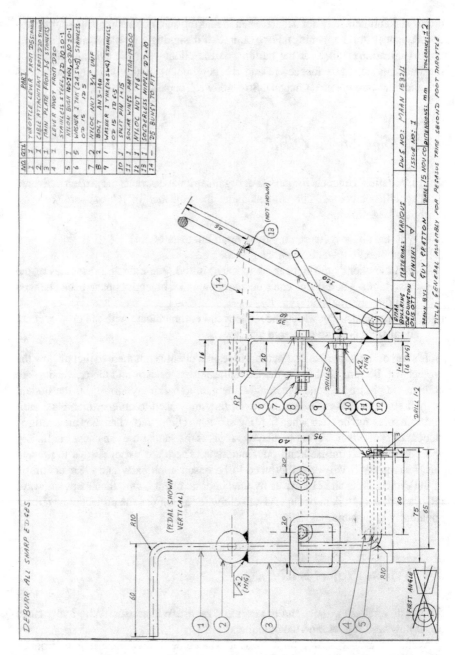

Fig. 1.2 Example of an Engineering Drawing (this is hand-drawn, which is still usually acceptable, but modern drawings are more likely to be computer generated)

prevent a student from failing, and are usually too busy to provide as much active support as they, and their student, might prefer.

Finally the client—well, they set the problem in the first place, and clearly want the answers that a good solution will bring. On the other hand, they might be a lot less interested in what marks the student gets, and even if they really want the student to do well, won't have as good understanding of what makes a good final report as the supervisor, nor (if they were paying attention!) as the student—although a good client may be able to provide additional support if it's warranted.

So, everybody owns the report, but only the student can put in the effort that really ensures success.

1.5 Dissertation Terminology

This section lists some of the main terminology that you may come across along the dissertation journey, and which are also used throughout this book. The authors have *mostly* used British English as a standard here and elsewhere in this book—but even within British English, do be aware that individual institutions and even individual departments or supervisors may use some specialist words in different ways.

Sometimes, you may need to define words, or groups of words, with a narrow meaning that matter for your specific project. If you do, ensure you do so clearly and early in your report in a similar manner to the way we have done here: in specialist

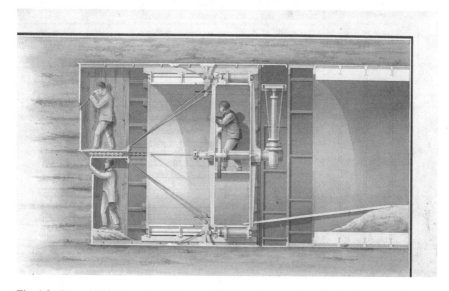

Fig. 1.3 Example of an Engineering Sketch (nineteenth century illustration of Brunel's Thames Tunnel). (https://cdn.antiquestradegazette.com/media/20656/web-brunel-tunnel-1.jpg)

technical writing, it is often helpful to clearly define your reader how you are using particular words.

A	Abstract	A short passage, normally at the start of a paper or report, which summarizes its contents. It is not the same as a conclusion, nor an introduction.
	Academic Journal	A journal (which may be printed, online, or most often both)—where research papers are published after having been subjected to expert peer review before publication. Sometimes referred to as a "peer-reviewed journal". This is often just abbreviated to "journal".
	Analysis	The process of using analytical (in engineering, usually mathematical or computational) tools to process and evaluate information
	Appendix	A section at the end of a paper, dissertation or report containing information which is self-contained and related to the content of the paper, but not essential to the understanding of the paper—therefore did not need to be incorporated in the main body of the document
	Archive	Somewhere that documents and data are stored securely for future reference. Historically, this would normally be in a paper/boxed form; nowadays, more likely a non-expiring website or institutional computer data store.
	Accuracy	The maximum difference between a measured value and an unknown true value
B	Bibliography	A list of related documentation, normally not directly called out from the text of an associated document
	Business Attire	The mode of dress and appearance that is considered normal for professionals within a workplace. This can vary significantly between locations, companies and so on, although note that generally the engineering profession tends to have more formal dress codes than academia.
C	Capstone Project	A substantial, non-trivial piece of coursework at the end of a course of study, which is devised to integrate several topics and showcase a breadth of learning from the course. This term is commonly used in North America as an alternative term for a dissertation.
	Carpet Plot	The representation of three or more dimensions of data in a two-dimensional format, through visually illusory further dimensions
	Citation	An entry within the text used to show a link to an external document (details of which will be listed more fully in the references list). Sometimes also called a "call-out"
	Client	A person or an organization who has defined the desired outcomes of a project and has specific interests in those outcomes being met

(continued)

(continued)

	Common Data or Common Variables	Associated with tabular or graphical presentation of data; this is the statement of common conditions that apply to everything in that presentation. It is included so that if the table or graph is used on its own, it is still clear what it's about.
	Conclusion	At the end of a paper or report, this is the statement of what the narrative content of the report led to
	Conference	A meeting where people who have carried out work or research give presentations of that work to colleagues
	Conference Paper	Similar to a journal paper, but normally shorter and has (sometimes) not been subject to formal peer review. Presented by one or more authors at an academic or professional conference. A paper may be accepted by conference organizers, which means they will publish it in the Conference Proceedings, but the authors may be invited to present the paper as a poster, rather than in person, as the slots for personal presentations at a conference are limited by time.
	Conference Presentation	A presentation at a conference, where one or more authors present their work to an audience. It may sometimes be accompanied by a Conference Paper (see also Poster)
	Conference Proceedings	This is a publication of all the papers presented at a particular conference. It should be understood that the contents are sometimes about work-in-progress, and often have not been subjected to peer review prior to publication. Some conference organizers instead provide collated abstracts
	Critical friend	Someone whose judgement you trust to provide you with a balanced appraisal of your work; similar to a "study-buddy" but might be a peer who is a little more advanced in their studies or education than you, or a family member already a professional engineer
	Critical Path Method (CPM) or Critical Path Analysis (CPA)	A method used to analyse the predicted duration and interconnectedness (contingency) of tasks in a plan, and thus predicts the total (minimum) duration of the plan and provides mechanisms for showing planning issues. Tasks not on the critical path of a project schedule are said to "float" between points in time. The concepts of CPA are embedded in other project management techniques such as PERT and Prince II
D	Data	Plural items of information
	Datum	A single item of information
	Definition	An explanation of how a particular term is being used within a document
	Derived Quantity	A value representing a property of something, which has been determined by performing calculations from measurements
	Dissertation	Term for the report related to a major student project (this term is not used in every institution), frequently used to refer to the project itself, often informally abbreviated to "disso"!
	Drawing	A scaled graphical presentation of a design, which includes such critical information as dimensions, tolerances, materials and manufacturing instructions (Fig. 1.2)

(continued)

(continued)

E	Error	The maximum difference between a measurement, or calculation of a quantity, and the (usually unknown) true value.
	Error Analysis	The process of estimating the differences between measured, estimated or calculated data, and the true values
	Error Bar	Symbology used on graphical representations of data to indicate the accuracy of the data
	Ethical Approval	A process whereby a research plan has been assessed against an ethical standard, normally by a panel of experts. It is usually mandatory for all research involving interaction with animals or human beings, but requirements for other research will vary with institution
	Examiner	A person, normally independent of the commissioning and supervision of a project, who will be assessing the project (normally through a combination of the written report and oral presentation) and assigning grades/marks
	Extenuating Circumstances	When personal circumstances outside the control of a student (e.g. illness or bereavement) adversely affect their academic performance; a university process whereby it is considered whether adverse personal circumstances should lead to special considerations in supervision, deadlines, marks, re-sits and so on. Generally, this is controlled by a central reviewing committee, and is not in the gift of individual academics. Procedures will usually be consistent within one university, but may vary between universities.
F	Feasibility Study	A process of determining whether something is possible and how it should be tackled: usually including evaluation of necessary time and resources, and technical risk. In the context of a dissertation, a feasibility study might be the start point of a project—or the whole project itself
G	Goal	Synonym for overall aim of a project and used in some universities/departments
H	Harvard	A specific system for calling out references from the text, sometimes referred to as the "Author, Year" system
	Hypothesis	A statement of how something might be, which is then tested. For example, we might hypothesize that the moon is made of cheese—but would need to take careful measurements, probably in situ, to confirm or deny that this is the case
I	Interview	A process where a researcher questions somebody to obtain information. Interviews are usually used to obtain qualitative data, in which case it is usual to use a structured or semi-structured format to ensure the same set of data is obtained in each interview.
	Introduction	The opening section of a piece of written academic work, setting the background and context
J	Journal Paper	A research document which has been subject, successfully, to formal peer review and, as a result, has been published in an academic journal

(continued)

(continued)

K	Keyword(s)	Terms used to highlight the most significant topics addressed in a piece of work, and used within indexing systems to assist future researchers in identifying what pieces of prior work they may wish to access and make use of
L	Lab Book	The document used to record observations, measurements and actions of experimental work, normally in chronological order. Historically, this would be a handwritten book of notes, but nowadays may equally be a computer document
	Learned Society	A national or international organization of professionals and academics, which exists to support the development of knowledge and professionalism in a particular field or sub-field
	Literature Review	A preparatory report, eventually integrated into the final report, which summarizes and analyses the existing published information on the topic of the dissertation
	Logbook	A record of actions, notes and ideas, normally in chronological order and often used to record and support career development or track the lifecycle of a project (somewhat similar to a lab book, but more generalized)
M	Measurement	A directly obtained value of a quantity: for example, the use of scales to determine the weight of an object, or of a rule to determine its length
	Meeting	An occasion where several people gather either in person or by electronic means to share information and draw conclusions about subsequent actions
	Metadata	Information about information ("data about data") that helps in its categorization, and contributes to its understanding and analysis
	Methodology	A series of actions by which a combination of experimentation, measurements and calculation, as well as analyses of data, is used to reach some conclusions
	Mock-Up	A representation of a proposed machine, design or structure which is visually representative of a proposed product, but which is not fully functional (see Prototype)
	Model (computer)	A computer program that aims to simulate a series of physical processes, built upon equations that aim to represent those processes and values representing conditions. Such a model is used to understand the processes or predict outcomes (e.g. to predict the failure loads, or failure modes of a structure under load.)
	Model (physical)	A form of mock-up, but more likely to be scaled, and less likely to be representative of functionality
N	Networking	The process by which people with common interests or intersecting objectives meet and interact with each other, so that they have future potential to support each others' goals
	Nomenclature	The section of a paper or report which defines what acronyms and symbols mean
O	Objective(s)	A statement of the reasons and hoped-for outcomes of a plan

(continued)

(continued)

	Opinion	A set of conclusions, built upon incomplete but best available evidence
	Oral Presentation	A presentation, possibly accompanied by visual aids, where somebody explains a position or body of work (see Viva Voce)
	Original Work	Work (physical, written, performed or electronic) of the stated author, and does not contain any significant content attributable to anybody else
P	Pilot Study	This is a scaled-down investigation to test whether methods of data collection and analysis are going to work before carrying out the main piece of research. Refinements to methods following a pilot study can greatly improve efficiency or accuracy
	Plagiarism	The use in a piece of work of material that isn't the work of the author(s) without clearly identifying the true source of that material
	Plan	A projection of the intended timeline and allocation of resources to deliver an outcome
	Poster	A document that explains a piece of work, in a form similar to, but simplified from that of a journal or conference paper—on a single side of paper typically around A2 to A0 size, designed to be displayed on a wall. [Examples of posters are shown in Sect. 16.1.]
	Precision	The numeric resolution to which a value is stated
	Primary Data	Original data obtained through observation, measurement or calculation. For example, from billing data you can calculate energy consumption for specific periods, the billing data being primary data for your analysis.
	Primary Source	A document or record that contains original unmodified observations or commentary. For example, a lab book containing original readings, or an account by an eyewitness of an event.
	Product Design Specification	A document describing in detail the characteristics, functionality and performance (intended) to be achieved by a product under design
	Professional Accreditation	A system of standardizing qualifications (and consequently courses of education and training) across a profession (e.g. engineering), usually administered within country (or region) and often mutually recognized internationally (although not always in the eyes of the law); for example, the designation PE in the USA has equivalence to CEng in the UK, but they are awarded locally
	Proposal	A statement prepared by somebody who wishes to do something (e.g. a student, a researcher or an engineering contractor) defining what they intend to do, how they will go about it, what resources are needed and why they should be resourced to do it
	Prototype	A working model of a product: the functionality might be incomplete compared with the intended final version of the product, but this kind of artefact is more developed than a mock-up
Q	Qualitative Data	Data about something which are non-numeric; for example, a survey of descriptive opinions, contents of interviews with stakeholders, drawings or photographs
	Quantitative Data	Data about something which is all defined in a numeric form

(continued)

(continued)

	Questionnaire	A paper or electronic form used for recording data provided normally by a single person
R	Recommendations	Recommendations come at the end of a paper, and are proposals for new work, projects or other changes that might usefully come as a result of the report's conclusions
	Report	A document which details a body of work, usually beginning with an introduction and finishing with conclusions and recommendations. The layout and internal structure may vary significantly, and often is required to follow specific guidelines.
	Reference(s)	References are a list of publications that are used within a document. Normally, the contents of a reference list are all specifically used in the new document, and called out from the text where they are used.
	Research Question	A question for which an answer is unknown but desired, and therefore, a body of research being carried out in an effort to answer the question. The research question can be manipulated into a statement, which is the principle aim of the project.
	Risk Assessment	A structured method by which the safety risks associated with an activity are assessed, quantified (typically as "low", "medium" or "high"), mitigating activities identified, and then the residual risk level formally accepted
S	Secondary Data	Data obtained from a published source, but which is not original data. For example, it could be a journal paper analysing previous primary data, or a secondary account of an event
	Secondary Source	A document which contains information, which is (fairly) authoritative, but is not the original source of the information and thus almost certainly includes a degree of filtering or interpretation. Most textbooks should be regarded as secondary sources
	Sketch	A graphical illustration of a system, design or idea which does not contain detailed dimensional, manufacturing or component identification information (Fig. 1.3)
	Stakeholder	Somebody who is not specifically interested *whether* a project succeeds, but has an interest in what its outcomes are. For example, neighbours of a building project do not necessarily have a vested interest in whether it is built—but do have strong interests in what it looks like, its impact on nearby parking and so on. Thus, they are stakeholders.
	Student	Somebody registered on a course of study
	Study-buddy	Someone undertaking the same (or a similar) course as you and who can work with you to support each other's learning process
	Supervisor	A member of academic staff who is responsible for overseeing a dissertation project
	Survey (of an item)	A process involving measurement and inspection of the physical dimensions and conditions of buildings, plant and equipment. This might be as simple as a visual inspection, including noting details on data plates, or as complex as using thermal imaging to detect relative temperature variations.

(continued)

(continued)

	Survey (of people)	A process whereby multiple people or organizations are each asked the same questions so that their answers can be compared and analsyed
T	Technical Memorandum (TM) or Technical Note (TN)	A (usually short) document that describes either current, or emerging, best practice in a discipline; often published by learned societies or trade associations, but the format might also be used within a particular business, especially multinational companies, to ensure consistency in their operation
	Test Plan	A formal document which describes how an engineering artefact will be subject to experimental evaluation
	Test Rig	A piece of equipment used to test an item, and obtain from those tests consistent experimental results
	Thematic Analysis	Analysing qualitative data so as to determine its meaning
	Theory	A testable set of principles laid down to explain a phenomenon or process that can potentially be used to predict future outcomes
	True Value	An abstract concept of what a value of a measurement actually is, rather than what it is read as (See Error)
V	Visual Survey	A non-invasive evaluation of something's form. For example, a visual survey of a building would include assessing what can be seen from the outside, number of windows, apparent condition of structure—but would not involve removing or penetrating any structure
	Viva (Voce)	An oral examination where a student presents and is questioned on their work by one or more examiners. This term may be used for dissertations, but is most commonly used for PhD examinations
W	Word Limit	The maximum number of words permitted in a document which is to be assessed. This normally is _not_ a target and may, or may not, include abstract, appendices, references and so on, depending upon local regulations. This may sometimes instead be expressed as a page limit.

Chapter 2
Objectives and Expectations

Abstract Students have three main objectives from the dissertation: they want the best grade, to use the dissertation to develop their graduate career and to learn as much as possible from the project. Supervisors want to see students succeed, but need to balance this with many other demands on their time, and can't stop a student from failing, only support them as well as possible. Clients are people or organizations supporting a project and hope to see useful outputs; they may be interested in recruiting graduates. Other stakeholders may also exist, and need identification and consideration.

© Springer Nature Switzerland AG 2020
P. Gratton and G. Gratton, *Achieving Success with the Engineering Dissertation*,
https://doi.org/10.1007/978-3-030-33192-4_2

2.1 The Student

2.1.1 The Objectives

If you want to graduate, of course you need to do well in your dissertation. Isn't that enough? Well no, not really: you want three things out of your dissertation:

(1) The best possible grade
(2) To produce something that'll help get a good job after graduation, not to mention
(3) To learn as much as possible, which you can use as you start your professional career.

We've seen too many students who think that it's all about the first of these: but the best students aim to get all the three right. In doing so, you'll probably get better grades as well.

2.1.2 What not to Expect

To achieve these things, what do you expect from everybody else? It's perhaps easier to say what you can't expect. You can't expect lenient marking saying that your work is better than it really is, you can't expect anybody else to do the work for you and you can't expect enormous resources handed to you on a plate—there will be dozens, probably hundreds of students doing their dissertations at the same time as you, so the resources need to be spread thinly.

2.1.3 What Actually to Expect

What you have every right to expect is straight and honest treatment by the university, your supervisor and anybody else involved in the project. This includes:

- You should expect to have been given sufficient opportunity to learn everything you need to learn to do well. This doesn't mean you can expect to be spoon-fed: think of it like a gym membership. The fees get you in the door, but you've still got a lot of work to do to get fit.
- You have a right to expect anybody who is supporting you to deliver what they've agreed, particularly meetings when agreed, lab space promised and access to information sources.
- You have a right to your work being fairly assessed, against the criteria that are published.
- You have a right to a <u>reasonable</u> amount of feedback en-route (but that doesn't mean absolutely everything you do being checked and corrected: to a large extent

on the dissertation you <u>are</u> on your own, or to put it more positively, being treated like the professional engineer you're trying to become).

Notice that nothing in there means that the university also won't let you fail—you can't expect that unfortunately. Not failing, ultimately is your expectation from yourself, not anybody else.

Producing a "beautiful" dissertation sounds an odd concept, but to most professional engineers (you know, the people who'll be interviewing you for your first graduate job!) a well-written and crafted technical report is a thing of beauty. To help you do this, you should hopefully be provided with good clear writing guidelines, and have been given lots of technical writing practice up to now. If you haven't—well, this book should help you to develop good practices in both, but talk to your supervisor or the dissertation module leader also about developing these skills—and you have every right to expect something, although how each university does that will vary. Looking for that help early is a good idea.

You'd also like to be given every opportunity to learn from this journey—and it's important whilst doing so to remember that the dissertation is as much about learning as it is about being evaluated. You do start this process knowing a lot about all of the constituent subjects, but you've probably never done a piece of self-directed work this big before, and a lot of the subjects leading to the dissertation: such as technical writing and project management have only been touched upon so far. Also you're likely to use several—perhaps all—of the subjects studied so far. So, you need to learn—and learn quickly how to use some subjects for real, and many subjects put together. You can expect *some* help from your university as you do that—certainly guidance from your supervisor, and some universities will provide specific lectures and learning material. At the same time, resources will still be limited, including your supervisor's time, so don't expect that to be handed to you on a plate: it'll need working at. Ignoring everything else, and treating the whole exercise as a pursuit of grades will disadvantage you in both the short and long term—don't do it.

2.2   The Supervisor

2.2.1   The Supervisor's Role

It's the supervisor's job to oversee dissertations, and as such they, of course, are going to try and do their best to provide as much support to students as they reasonably can. However, it's important to remember that for any academic, this is part of a complex juggling act that is trying to balance demands upon their time in preparing and delivering lectures, research, contributing to management of their department, and writing books and papers. This is, of course, why so many supervisors will set or mainly agree to supervise projects which align closely with their other work (Fig. 2.1).

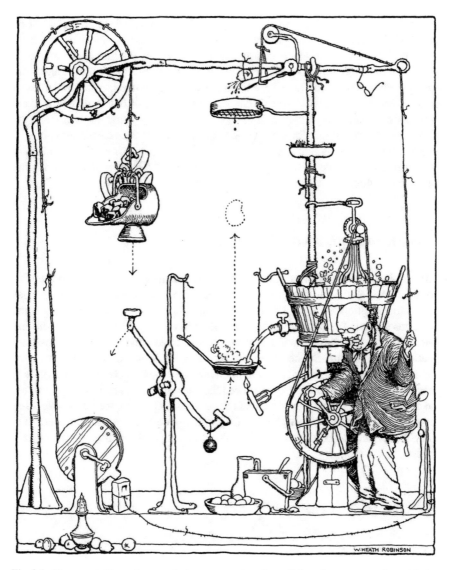

Fig. 2.1 You may not be quite sure what your supervisor does all day—but you can rely on it being important to them, and they'll be very busy doing it (Mr William Heath Robinson)

2.2.2 *Relationship with the Student*

While working with a dissertation supervisor, the single most important thing that anybody, especially a student, must respect is their time. The supervisor has a right, therefore, to expect that their time is used as efficiently as possible. From a student's viewpoint, that means above all else always turning up to meetings prepared and on time and always endeavouring to meet deadlines.

By definition, a supervisor is clearly experienced—they did their own projects as a student—through the PhD which is in many ways simply a much bigger and more complex dissertation, and are likely to have supervised dozens, if not hundreds of projects. So, they also expect to be listened to—their advice on how to do well is likely to be good. Clearly, they aren't infallible, but if somebody disagrees with them, the supervisor also has a right for those disagreements to be phrased clearly and politely, and if a student in particular chooses not to take their supervisor's advice (and every student has that right) then the supervisor expects to get a clear explanation of what the student's doing, and why.

2.2.3 Relationship with the Client

If the project client is somebody other than the student or supervisor, then the relationship between the supervisor and the client is more complex. The supervisor has agreed to a relationship with the client because of a mutual interest in seeing some problems solved. The student to some extent is in the middle of this, and will find that such a project limits their flexibility. It also makes the supervisor-and-client relationship important. So the supervisor in such cases looks for reliability from the client—that promised information and input into the project is always as complete as possible and on time, whilst they also need the client to respect that the supervisor is the expert on how to manage the student's work. In such cases, they're also looking for a respect from the student that the project is <u>not</u> entirely the student's, and that some things need to be done in particular ways to hopefully deliver on the client's needs.

2.3 The Project Client

2.3.1 Who Is the Client?

The client can be one of many types of people: it could be the student or supervisor, it could be somebody else in the university such as one of the leaders of an ongoing research project, or it could be a person or organization outside the university—for example, an engineering company or charity. What most clearly defines the client is that they want a problem to be solved. By providing support to someone's dissertation project, the client has every right to expect the student and their supervisor to at least make a reasonable stab at solving the problem. At the same time, they need to accept that the student and their supervisor are experts in their particular field of engineering, and also have particular ways of working, laid down by academic rules and best practices. So the client can't expect to dictate *how* a project is run.

To get that right, then the client would be advised to make sure they've got straight with the supervisor and student statements about who is doing what and when, and what inputs they do, and don't have, into the project. There's no harm in getting that in writing—although hopefully just a discussion or a few confirmatory emails should be enough.

2.3.2 Client Expectations

At the same time, there are things that the client should expect as a matter of course—regular updates about the progress of the project, a copy of the final report and to be informed promptly if anything goes wrong with the progress.

Something that is worth discussing between the client, supervisor and student is *Intellectual Property*—if the project generates IP, who owns it, and who can be involved in exploiting that IP? Different people (and the university as an organization) may have different views on that—and almost always the best way to deal with this is to discuss it up front, and if necessary get an agreement down in writing (do this plenty of time, as, especially if lawyers get involved, it can all get a bit slow).

These issues are often dealt with in a non-disclosure agreement (NDA). Clients will often use such documents to limit how much innovation within their business can be talked about, outside their organization. This can present a challenge to a student, who has to devise a report that demonstrates the learning they have gained personally in doing the project, without giving away information which the client may consider commercially confidential. Ultimately, the onus is on the supervisor (as the experienced academic) to ensure that the report is composed in such a way that it can be examined by any member of the university and ultimately archived in the university's repository (effectively publishing the document).

Project reports that require limited access by named examiners who have each signed an NDA, and have to be stored secretly for a limited or unlimited future should be avoided. Save secret research for your PhD!

2.3.3 Future Employment

Is the client interested in possibly employing the student after they graduate? Is the student potentially hoping for a graduate job with the client? There's absolutely nothing wrong with such expectations—but it's a good idea to discuss it, and often the client is best placed to lead that discussion. They shouldn't necessarily expect to hear what they hope for when opening a discussion like that—but at least it gets this out in the open. It's likely that the most a student can hope for, particularly with larger organizations, is a fast-track onto a shortlist for interview—smaller organizations *may* be more flexible.

2.4 Other Stakeholders

2.4.1 Identifying Stakeholders

A stakeholder is somebody who cares, or has a vested interest in how a project comes out. This isn't the same as wanting the project to succeed—although they might, nor wanting it to fail (although that can sometimes be true too). In the broader engineering world, it's very important to identify and engage with stakeholders; for example, the people who live near to a planned new building or around an airport gaining a new runway, patients who may hope to benefit from a new medical device, pilots who will be flying a new aeroplane or businesses that will be using a software upgrade. Some stakeholders might not want a successful outcome—fraudsters might not appreciate new and more secure financial security measures, or motor factors might not appreciate new tyres that last twice as long at the same price: but they still have a strong interest in outcomes. None of these people are the client, supplier or a direct customer—but they have a *stake* in how the project comes out.

It's always important in any engineering project to know who the stakeholders are, and their needs and interests. With a dissertation project, this is less true, only because as a project it's of far smaller scope than most real-world engineering projects: but it's still important to know who the stakeholders are, and why they matter.

Of course, the student, their supervisor and the client are all stakeholders—that goes without saying, and we've already analysed their interests. But others may exist—here are some examples:

- A group of disabled people who will use an ability aid being designed for a client, who is a support charity
- Young children who will use teaching aids being designed for a school
- The people who will occupy a building for which you're designing an improved piece of climate control software
- The lab technician who'll have to manage a piece of measuring equipment being designed for a research project
- Collaborators of your supervisor, who are acting as clients for a piece of basic research.

In discussion with your supervisor and client, it's important that you identify your main stakeholders (don't get silly about it—remember this is ultimately a learning project so if you can come up with a list of 20, it may be best to agree just consult with the most important two or three; on the other hand if you can't identify any, your supervisor might ask you to make one up, for the purposes of the learning exercise).

2.4.2 Stakeholder Relationships and Communication

Once you've identified your stakeholders, you need to identify what matters to them. Ideally, you can do this by talking to them, perhaps through an interview or a carefully constructed questionnaire. Sometimes this isn't so easy—if stakeholders are hostile to the project, unable to respond themselves, or for project reasons are fictional, then you'll need to be more creative. This might include interviewing experts who have dealt with such people before, use of press reports, or talking to somebody elsewhere but with similar interests. The range of scenarios here is too wide to be proscriptive in this book—it will need to be discussed between the student and their supervisor.

2.4.3 Ethical Approval

If you are going to approach stakeholders, whether to collect personal information or to canvas opinion, then BEFORE you attempt to make contact, you should obtain ethical approval from your institution. Usually, your university's intranet will provide information and guidance about how to go about this. If you are unsure, then speak with your supervisor. Most universities will refuse to examine work that has been submitted for which ethical approval has not been agreed. This is done not only to protect the reputation of the institution, but also to protect students from placing themselves in dangerous situations.

Once this has been done, it becomes possible to consider stakeholder "issues" and responses in the construction of the engineering project that is leading to the eventual dissertation report.

Chapter 3
How to Be the Student Who Achieves Success

Abstract Success with the dissertation starts long before the project is assigned. Students who wish to do well need to develop excellent subject knowledge, written language skills (including through targeted leisure reading) and have allied themselves with suitable student or professional bodies. Ensuring a good workspace (both physical and electronic) and managing time and personal attributes are important, as is establishing good working relationships with critical people such as administrators, technicians and key academics.

© Springer Nature Switzerland AG 2020
P. Gratton and G. Gratton, *Achieving Success with the Engineering Dissertation*,
https://doi.org/10.1007/978-3-030-33192-4_3

3.1 Defining Success

We've used the grand term "Achieving Success with the Engineering Dissertation" for this book what do we mean by that? First, you want the best grade available; secondly, you want to use the dissertation as a learning experience to make you the best engineer (or engineering researcher) that you can be at this stage of your career and thirdly that you want to use your dissertation to help you enter your career of choice after graduation with an upward trajectory. These three objectives are firmly linked and shouldn't be treated in isolation from each other.

This chapter will try to lead you through all three objectives—but is primarily about the period before you start your dissertation: a great way to fail is only to start thinking about the whole process once you've had your topic approved.

At the end of the chapter we've included a checklist for you to work through and tick-off as part of your preparations. Of course, some of the details here will depend upon your topic, country and local assessment regime—but the basic principles to achieve success are universal.

3.2 Before You Start the Project

Knowing your subject
Hopefully, you have an idea, from the year before the dissertation year, what broad topics you want to pursue—whether that is orbital mechanics, mechanism design or microelectronics. Clearly, a supervisor is more likely to accept you as their project student not only if you have good grades in the right subjects but also if you want to impress them, an oral examination panel and a future job interview panel—then you need to have a level of understanding far beyond just the ability to pass exams. So, make sure both your breadth and depth of understanding of all your major subjects are good, and that you're well able to discuss them.

The dissertation itself will almost certainly be on some narrower sub-topic (or combination of topics), and may well be in one that you have not yet studied at all. **Start reading around your topic early**, and keep reading (also keep a list of what you've read and where it is located). You'll be required to include a review of current literature in your project proposal, but there is nothing to stop you completely revising this and adding more recent publications before you submit your final report. Displaying this level of awareness of your field is one thing that differentiates a really good dissertation report. Clearly, when we say "read" here, we are also including attending extra lectures, visiting museums, talking to staff or more advanced students, listening to podcasts, watching online videos and so on.

Knowing your language
We're assuming here that you will be writing your dissertation in English language. As an engineer, your studies and work have been primarily about science, mathematics, and explaining concepts with numbers, diagrams and equations. The dissertation, however, whilst incorporating all of these things, will be primarily a work of written English—and probably the longest single written work you have ever produced. So, you need to be very good at both reading and writing clear, concise English.

A start to this is, as early as possible, to direct your leisure reading. There are a number of professional engineers and scientists who have become successful writers—and whilst hopefully enjoying their writing, you can develop a good intuitive understanding of good, clear written English such as you need to be able to write yourself in your dissertation. We've suggested some writers in Table 3.1, although you may find others for yourself and if not writing in English there may be better alternatives. Needless to say, whilst many of these authors' books have *also* been made into movies, some extremely good, that is not an acceptable alternative to reading the books!

Join the right clubs
Your field will have one or more local, national or international societies dedicated to it; these societies almost invariably have student or young-member chapters or groups. The range of these is extremely broad from the *Royal Aeronautical Society*, via *The Computer Society of India* to the *American Society for Artificial Internal Organs*. Student membership is almost invariably very inexpensive, and sometimes free—as is attendance at local meetings and larger society conferences. The larger

Table 3.1 Some suggested sources of leisure reading

The writer	Their background	Some suggested books
Arthur C. Clarke	Electronic engineer & physicist	Clarke, A. C. (1957) *Tales from the White Hart: a sparkling collection of scientific tall stories*: Ballantine Books Clarke, A. C. (1963) *Glide Path*, London, UK: Harcourt Brace
Nevil Shute	Aeronautical Engineer	Shute, N. (1954) *Slide Rule: Autobiography of an Engineer*, London: William Heinemann Ltd Shute, N. (1948) *No Highway*, London: William Heinemann Ltd
Homer H. Hickam Jr.	Aerospace engineer	Hickam, H. H. (1998) *Rocket Boys*, Delacourt Press [This book has also been published under the title "October Sky"]
Simon Singh	Physicist	Singh, S (1999) *The Code Book*, Fourth Estate

societies will have physical libraries, and many now will either have online library access, or some form of online papers database.

Which is the most appropriate society for you? In the broadest sense, that should be obvious by what your lecturers and professors are members of, and which organizations hold meetings in your campus. There may, however, be narrower society allegiances that you want to consider—for example, a British-based Aeronautical Engineer is most likely want to join the Royal Aeronautical Society, but if their project is related to flight testing, then there may be value in also taking out student membership of the USA-based *Society of Flight Test Engineers*. The most appropriate society may not always be headquartered locally, or even in the same country as you.

What should be clear, however, is that not being a member of at least one appropriate engineering institution, by the end of the year before you start your dissertation, would be foolish. And having joined them, attend local branch meetings, network with the people who attend, know your way around the resources on their websites, and whether it's feasible to directly access their library.

This is a habit which should also be carried on after graduation of course. Networking through your learned society is an essential component of establishing your post graduation career. Not just in finding opportunities for jobs, but connections that help you do those jobs. Many institutions also offer speaking or project competitions aimed specifically at young members that may provide opportunities to either provide direction to, or exploit your dissertation—we discuss this more in Appendix C at the end of the book.

Warn your friends and family

The dissertation is a very big task, and if you expect to do it well just working the hours nominally assigned to it, typically around 12–15 h per week, the results will be mediocre. Even if you get good grades this way, you won't get the full learning and career development benefit. Of course, the final year of an engineering degree is extremely demanding anyhow; but the dissertation will be the single most demanding component of it. Life is much easier if you live on campus with other students in the same year and similar courses, but everybody needs to know that you will be stressed, working long hours and regularly obsessed with the latest project difficulty. Don't let yourself get sucked into excessive family and social commitments that will interfere with you achieving your best possible performance. If they care about you, they'll understand when you explain your need to be selfish for a while.

The same is true of employers—many students are either part time, or have part-timetime jobs that are essential to their financial survival. This can't be avoided, but as with family, ensure that your boss knows that you will be entering a period of very high workload, and establish realistic expectations of the hours that they can expect you to work. Looking on the bright side, dissertation students are unlikely to have the time to party much this year, so won't need *quite* so much money.

Don't leave preparation too late
If you have only just started to read this after being allocated your project and been accepted by a supervisor, and have done no preparation so far—you are on the back foot. We hope you didn't do this—but if you did, the right answer is to start putting effort **now** into catching up, not ignoring these essential pieces of preparation.

3.3   As You Start the Project

Sort your workspace out
You have two workspaces in your dissertation: the physical and the electronic.

Your physical workspace is *where* you'll locate yourself as you work on the project. This doesn't need to be anything special—a table in the corner of your bedroom where you can plug in a laptop and have space to stack up some books and notes may be all you need; some people need silence, some people need their favourite music, some people work best with others around them—the Nobel prize winning physicist Richard Feynman did much of his best work on a table at the back of a strip club (almost certainly beyond both wisdom and budget for most undergraduates, but it worked for him[1]). If you have your own family, you might need a remote or lockable and childproof study—even if it's a temporary re-use of a broom cupboard. The important factor here is to know what you require, and ensure you have it (Fig. 3.1).

The electronic workspace is more straightforward—you need hard disc space, whether that is on your own laptop, Cloud-based, a portable drive or whatever is your preferred technology, and without fail METHODS IN PLACE FOR ENSURING IT IS ALWAYS BACKED UP. You should also make sure that you have working, and know your way around any software you'll be using, online library access, and have installed or bookmarked tools that you will need such as Web of Science, Google Scholar, the NACA Online reports server, Unpaywall and so on. Discuss the resources you'll need with your supervisor, who will have a very good idea of what's appropriate for the sort of project that you will be doing.

[If applicable] Get to know your team
Some projects are set as team-projects. If there is a team for your dissertation project, or there are other students working on linked projects, make sure you have met all team members; get to know everyone's strengths and weaknesses (as a team, you might even consider using psychometric tests, like the ones some recruiters use in assessment centres); and agree what your communication strategy is going to be—especially allow enough time for face-to-face meetings.

---

[1] Ottaviani & Myrick, *Feynman*, First Second Books 2011.

Fig. 3.1 The right physical workspace will be different for different people

Read the brief!
Make sure you know what you have to do, and what the deadlines are. In project management terms, these are the non-negotiable milestones. It's a good idea to put these on the wall calendar that you will see every day.

Keep enough time committed
Your university might give you guidelines about how many hours to dedicate to your project, but this will be a total or average and you have to decide from week-to-week how long you are going to work on it. Students who succeed in their dissertations are on top of the work, but usually have a life outside it. To do really well, they need to make the most of opportunities aside from sleeping, eating and seeing loved ones (very important), attending lectures and labs, completing required assignment exercises, participating in sports, working for money, volunteering and going to social events. It helps if you have friends who are in the same position as you, but, if you are a postgraduate, mature or a distance-learning student, you might need to try a little harder to find a "study-buddy" or "critical friend" (these terms are explained elsewhere in this book).

Consider your own attitude

Success is a matter of attitude and opportunity. Attitude is an inner, personal quality that is going to manifest itself as your enthusiasm for the topic of your project. The best dissertations are usually authored by people who are *excited* by the topic of their project, and who have the skill to convey this passion to their examiners. At the same time, you need to have the humility to constantly accept guidance from "those who have gone before" such as your supervisor, more advanced students, and subject matter experts.

Know what you really want to achieve

Know what you aspire to gain from your dissertation. If you want to launch your career as a researcher, then use this opportunity to familiarize yourself with leading edge of your field of interest. If you want to join a design consultancy, then find out what particular engineering challenges their projects involve. If you want to get into a particular sector of industry, make sure you are up-to-date on reading the trade literature and business pages to know what is of current concern (e.g. artificial intelligence, climate change, energy security, digital security) and what is the next "big thing". Having joined the local branch of your professional institution, never be afraid of asking for an introduction (just ask the institution rep or local organizer)— sharing knowledge is what these meetings are for, both formally and informally. Networking is also a great way to find industry support or (even!) sponsorship. You might even find someone with a project idea, who is looking for your knowledge and skills.

Get to know the right people

There's an old saying: "it's not what you know, it's who you know". This is never completely true, but there's always an element of truth to it. There's no need to go overboard with this, and of course, you probably already know your supervisor quite well. But consider who else you at the very least need to know who they are, where they are based on campus, their working hours and how they prefer to be contacted? This is likely to include:

– The departmental administrator.
– The technicians for any labs you're likely to need to use.
– The subject librarian in the departmental or university library.
– The project client (if there is one).
– Any research students, post-docs or academic staff working in the area of your dissertation.

It is likely that a project that does not take adequate account of the people you need to work with, will struggle. Working with people in this way is also valuable practice for your postgraduate career, where it will always be extremely important. Engineering is a team sport.

It's also very worthwhile having a "study-buddy"—ideally another student going through the same as you are. This gives you both the opportunity to check each other's work, discuss thorny working problems, act as a proofreader, review presentations before they are judged and so on. This isn't essential, and many students manage without one – but most people who have had such a relationship find that it helps a lot.

3.4 While You Are Doing the Project

Stay in regular contact with your supervisor, and ask lots of pertinent questions. Most universities let project students initiate supervisory meetings. It's good practice to set up regular meetings at the start of the project, but you don't need to feel that you have to "save up" questions until then. If it's the sort of question that only your supervisor can answer, then sometimes a short email exchange between meetings can resolve the query. If you can't keep an appointment, or you have no progress to report, or are struggling, then let your supervisor know as soon as you can.

If your project has a client (other than your supervisor), then all the advice above applies equally. Be aware that their time is precious (it really does cost money), and they will expect you to conduct yourself professionally in a meeting [this *probably* doesn't go as far as wearing a suit – but check]. Talk to your supervisor about this, if you are not sure what is needed. If you can't keep the appointment, let them know well-ahead of time, as nothing prejudices their opinion against you like being stood-up! We talk about this more in Chap. 8.

3.5 Checklist

The following checklist is for you to use as your project progresses. The questions are in no particular order of priority, but have been grouped around themes. Some questions might be irrelevant to your project, so you can put a line through them. Some questions might need reviewing as your project progresses. Use your answers to this checklist to help you construct your Gantt chart. It's written in the first person, as these are questions you must ask yourself, and answer honestly.

Personal attributes

☐	I have genuine enthusiasm for the topic of my project
☐	I have sufficient time to commit to the project
☐	I have the subject knowledge necessary to start the project
☐	I've read and understand the project brief / guidelines / marking criteria
☐	I have the confidence to admit what I don't know and ask questions
☐	My networking skills are appropriate to ensure I meet the professional contacts that I need to enhance my topic understanding
☐	I have the communication skills to conduct meetings
☐	I can concentrate to read and make outline notes for my project
☐	I have the diagnostic ability to breakdown a problem into its component parts
☐	I have the organizational skills to prioritize the various aspects of the project (and all the other things in life!)
☐	I have the intellectual skills to synthesize ideas together logically
☐	My written English is of publishable standard
☐	My personal presentation skills are honed to provide an informative, engaging, live, oral presentation.

Working Environment

☐	I have a preferred workspace, which I have control over
☐	I have somewhere where I can write comfortably
☐	I own or have the login rights to the software I need
☐	I have electronic storage capacity to save my work in
☐	I've got data back-up to at least one additional place
☐	I have laboratory or workshop access for experimental work
☐	I know what times I have access to the lab / workshop
☐	Any safety training has been completed and signed-off so that I am allowed to use the facilities I need.
☐	I know how to access specialist libraries
☐	I've got links and access (login) to required data sources
☐	I have a suitable place to rehearse the oral presentation
☐	I have access to the university library and borrowing rights (no outstanding fines).

Relationships

☐	I have a "study-buddy" or "critical friend" who I trust that I can discuss my project with
☐	I've got a supervisor
☐	I've established a good working relationship with my supervisor
☐	[If applicable] My team mates have met and exchanged contact details
☐	[If applicable] Our team has established a group communication strategy (email list, Snap-chat, WhatsApp, Yahoo group, etc)
☐	If there's a client, I have established a working relationship with them
☐	I've met with the technicians who look after the laboratory / workshop where I'll be working
☐	I've got the contact details of the Subject Librarian for my course
☐	I know who the subject matter experts are and how to contact them
☐	My family and friends know how much time and effort I must dedicate to the project
☐	I've co-ordinated study time needed with my work.

Equipment

☐	I have a means of recording meetings that can easily be reviewed and annotated.
☐	The budget for my project has been agreed with my supervisor / client / departmental manager
☐	I own or have use of (a) suitable computer(s)
☐	I have /have access to the software I need
☐	I have /have access to any equipment (personal tools) I need.

Process

☐	I know the deadline for final report and any interim reports.
☐	I've got the date, time and venue of the oral examination
☐	I have meetings booked with my supervisor on the following dates and times:
☐	I've agreed number and frequency of meetings with my client /industry sponsor

3.6   Suggestions for Further Reading

Engineering Council UK (2014) *The Accreditation of Higher Education Programmes*, 3rd Ed., London: EC UK

Syed, M. (2011) *Bounce: The Myth of Talent and the Power of Practice*, London: Fourth Estate

Chapter 4
Selecting the Project

Abstract It is critical to everybody to make a good match between the project and project team, especially the project / student combination. The student of course also wants to ensure they are working with a supervisor and client who will provide good support and with whom they can work well. Supervisors will particularly wish to try and achieve good synergies with their research objectives and ongoing industry relationships. Clients will want confidence that they are working with a university, student and supervisor who will achieve good project outcomes. All must endeavour to achieve good team working and communications.

© Springer Nature Switzerland AG 2020
P. Gratton and G. Gratton, *Achieving Success with the Engineering Dissertation*,
https://doi.org/10.1007/978-3-030-33192-4_4

4.1 Advice to Students

4.1.1 Owning Your Learning

The engineering dissertation, whether at undergraduate or Masters level, is life-changing! It can be a topic in which you may find your future PhD , or the basis of your first job, or people with whom you develop significant relationships and so on, more fundamentally, it can be what stands between you and gaining your degree. So, choose your topic wisely!

Some university engineering departments expect, or require, students to propose their own projects and liaise with staff to persuade someone to supervise them. The alternative is that supervisors will propose projects and students select them. The dissertation project is a great opportunity to learn about something you are really passionate about, or fascinated in, and you are most likely to achieve success with the dissertation if you own the experience. This means selecting the right topic for you and finding the right support—definitely support within your university, but perhaps from outside it too.

Whether your project is your own idea, or from an idea given to you by a manager in the workplace, or from a project title you saw offered by one of your professors, you need to "own" it and take control of your own destiny.

If you have a choice, then select a project on a topic you like—preferably one that you are passionate about because you are going to "live" it for a long period of time. How much time you need to spend on your project may vary from course to university course, but it is likely that a third to half of the study time in your academic year is going to be spent on this major project. It will—and it should—come to dominate your life. So, prepare yourself (and your family, and those you share your life with), enjoy it and get the most out of it.

We'll talk about the amount of time you'll need to allocate to the dissertation project later in this book.

4.1.2 Selecting the Project and Supervisor

If you have a free-hand to choose a subject, then select one that will aid you in your future as an engineer. If you have not yet had paid-work experience in an engineering environment, the dissertation project can be an opportunity to make contacts in industry that will give both an insight into their workplace and enable you to introduce yourself to people who might either make you a job offer, or provide you with an introduction to someone who will. At the very least, you should later be able to ask them in future to endorse you with a character reference.

A dissertation project will always require you to work with a supervisor, who is a member of academic staff of your university. Arrangements will vary from place to place, for example, it might be that a supervisor heads-up a team of research fellows and PhD students, who are also available to support you. Occasionally, supervision of a project might be controlled by a committee of two or more academics. The supervisor role is one of providing guidance, especially with regard to meeting the academic requirements of the project. This is particularly pertinent when you are developing your own project idea, or one that you have brought in from industry. The supervisor is NOT there to teach you how to do the project! They can act as your mentor, so they can help you to find solutions, but they should not be telling you explicitly what to do every step of the way.

If you have the opportunity to choose your supervisor, then select someone you respect and think you can get along with. You need to be able to have honest conversations when you don't understand something, and you need to be prepared to be challenged by them. If you are afraid to be candid with your supervisor when you hit a problem during the project, then you are unlikely to be successful. Likewise, if you are not prepared to raise to the challenges they set you, your project is unlikely to be a success. Choose a supervisor you like, but not necessarily the person who will give you an easy ride.

In all likelihood, you would have already decided to specialize in a particular engineering discipline before you select your dissertation project. However, there is scope within disciplines to select a project that is dominant in one subject rather than another—for example, thermodynamics or strength of materials. If you are weighing the options over a range of possible projects, then choose a topic in an subject you already have a good track record—consider your subject results for previous years, and where you enjoyed the greatest success.

If potential supervisors in your department advertise projects, review what they are offering. If you are thinking of continuing your studies to a Ph.D., then this is often the way to find a studentship and supervisor. Of course, it is quite likely that you are in competition with your classmates in this kind of system. Once all applications for projects are "in", some kind of selection process will be in place for popular projects. This might be as simple as "first applicant gets it", or slightly more discriminating—student with the highest results in the course so far gets it— or you might find yourself being called for interview, or asked for a proposal. If you do have an interview, your prospective supervisor will probably want you to demonstrate your understanding of the investigation to be undertaken. For example, if a feasibility study has been set, do you understand what is NOT required?

Once you are allocated a project—either by a supervisor within the university or a manager within your workplace—be careful to review the "brief" carefully. Don't be tempted to change the aim of the project, or deviate from it, without checking with the "owner" (supervisor or manager). Of all the things that can go wrong in a dissertation, answering the wrong question is one clear way of avoiding success!

4.2 Advice to Supervisors

4.2.1 The Value of Student Projects to Student Engagement

Supervising dissertations, especially undergraduate ones, is usually one of the most rewarding aspects of university-level teaching. The one-to-one nature of the arrangement permits you to develop a relationship with an emerging engineer, so that you can see up-close the development of their knowledge and confidence of handling difficult technical problems. For many of us, we'll forget who taught us differential calculus but we're unlikely to forget our dissertation project supervisor.

4.2.2 Fitting in a Student Project with Your Research

If you have been asked to supervise a student dissertation project, or you have been asked to propose topics for students to select, then it is recommended you walk a line between keeping topics close to your research and teaching area and interests, and allowing yourself to be flexible enough to supervise things that are marginal to your personal interests. Today's students may be tomorrow's collaborators.

There are institutions that require dissertation projects to be written up such that the final report is formatted as if for submission to a journal, and some are actually submitted. It could be viewed that successfully completing the peer-review process is the ultimate evidence that it is worthy of an award at the highest grade! It could also be considered a vindication of the time you have spent in the supervision process.

Occasionally, you might supervise a project, the output of which feeds into a much larger research project that you are engaged with. You will not always know at the start of a student project whether they are going to produce anything worthy of publication. It is slightly more likely that they will generate some data that you can use elsewhere, and you can decide for yourself how significant this is to include the student-engineer as a collaborator. One of our professorial colleagues counsels low expectations of student dissertations, and reserves the right to be pleasantly surprised when they are successful!

We can boast some success in this matter; in 2014, the Bronze Medal of the Royal Aeronautical Society was awarded to Gratton, Hoff, Rahman, Harbour, Williams and Bromfield for their paper, "Evaluating a set of stall recovery actions for single engine light aeroplanes", which appeared in The Aeronautical Journal that year.[1] At the time of research, Messrs Rahman, Harbour and Williams were undergraduates studying Aviation Engineering with Pilot Studies at Brunel University London, who had completed individual projects, having flown as flight-test observers during the program.

---

[1]Gratton, G. B., Hoff, R. I., Rahman, A., Harbour, C., Williams, S., & Bromfield, M. (2014). Evaluating a set of stall recovery actions for single engine light aeroplanes. *The Aeronautical Journal*, *118*(1203), 461–484.

4.2.3 Managing Expectations—Differences Between Undergraduate and Postgraduate Projects

Managing expectation can help smooth the way in the student–supervisor relationship. You are obviously aware of what the learning outcomes are for the dissertation project. Your student may not be familiar with them, and may come to understand them only on completion of the project. As with the role of supervisor to the PhD researcher, the role of dissertation supervisor is more like that of a guide, or mentor, than of teacher. As you know, the differences are subtle, but the deepest learning occurs when the student figures something out for themselves. Your role is to facilitate the opportunity to learn from the project experience.

When you set, or define, a project, you need to have appropriate expectations for the level of degree, for which the dissertation is being done—but allow stretch, if the student shows capabilities. A common question asked is whether an undergraduate dissertation *has* to be innovative. Whilst we all acknowledge that the final report must be unique (and free of plagiarism), so the correct answer is *yes*, it should be acknowledged that innovation can be achieved in the tiniest of increments. Generally, not all engineering dissertations need to be limited to what are essentially *research* projects (although we're aware that some institutions insist on this). So, there is scope to steer a project student to *enhance* a previously carried out project.

One of the key discriminating factors between undergraduate and postgraduate dissertations is uncertainty. One of the essential learning outcomes for the project student is how to deal with uncertainties, and especially how to deal with the sudden need to revise the scope of their project. For undergraduates, who have no clear idea of their own capabilities and capacity, they tend to be either over- or under-ambitious for their project. This is often where managing expectations needs to be practical. A colleague suggested that one of the means of doing this is to limit the number of objectives in a project. Inexperienced undergraduates, who have about six months in which to carry out their project, probably should only attempt to tackle three objectives, all of which should clearly target one aim of the project. Slightly more experienced postgraduates might be encouraged to tackle six objectives in support of a single aim for an MSc project. If progress falters, then the student should be encouraged to reign in their ambitions, and cut off an objective. If progress is made quickly, they might be encouraged to consider adding an objective to their schema.

4.2.4 Collaborating with Industry

Supervising dissertations sometimes provides opportunities to work with industry. Sometimes, really engaged students will bring ideas for projects from their workplace (placement or permanent), but they need supervision to ensure that their project report meets the academic needs of their degree, as well as provides useful solutions to industry. Alternatively, you might be approached by someone from industry, who is seeking some help with a problem—this might be one of your alumni, who has some idea of how your university carries out projects. The terminology we have used in this book to describe the industry person who gets involved with a student project is project client. Your own institution might refer them as project sponsor or donor. The next section of this book (Sect. 4.3) is about helping them manage their expectations of collaborating with the university.

If you are looking for industry projects to collaborate in, then an obviously place to seek assistance is your university's "research development" office (other titles may be in use). It is the role of this office to liaise between academia and industry, and make introductions. The people employed there are "super-networkers", who tap into industry contacts, funding bodies and organizations like the Chamber of Commerce (local) or governmental departments of trade (national) to make connections.

In the UK, the principle organization for developing alliances between academia and commerce is the National Centre for Universities and Business (www.ncub.org.uk) which administers a database called "konfer" to encourage collaborative projects between its constituent members. Many other sub-field-specific databases also exist, but are most usually within the learned societies.

Collaboration with industry-based partners can be achieved through a variety of routes. Whichever way it comes about, you might need to consider whether the work requires a Non-Disclosure Agreement (NDA), and whether fees are chargeable, or will there be a reciprocal agreement (expertise traded for use of equipment or materials). If you have concerns of this nature then your organization's "research development" office is likely to have a legal expert to provide appropriate advice.

4.2.5 The Project Acquisition Process—Selecting a Suitable Student

How dissertation projects are acquired by supervisors will vary from institution to institution. There is usually some compromise between "selection" and "allocation". In some places, a "Hunter Paradigm" might predominate. This is where it is entirely left to students to have an idea, and track-down a willing supervisor. This method favours highly motivated students; however, problems arise when popular staff becomes inundated with requests, whilst less popular lecturers, or the ones who are lecturing in topics perceived as "difficult" by students, are left with no projects.

Another means to acquiring projects might be referred to as the "Candidate Selection Paradigm". A student can only approach several supervisors and discuss with each a prospective project, perhaps one advertised by the supervisor or occasionally one devised by the student. The supervisor interviews a number of candidates for a number of projects within a defined timescale, at the end of which they agree who they'll take on. The students can sign-up with the supervisor who has listed them, but if they have managed to run around and get on several lists, they might be in the position of having choices of projects to sign up to! This method can lead to acrimony on the part of the supervisor, if they get rejected by all their preferred candidates. Perhaps this method provides students and staff with valuable interview practice; alternatively, it could be seen as an ineffective use of time. It is certainly a messy arrangement to administer.

Another variation might be called the "Voting Method". In this case, supervisors are asked to suggest projects, and a list is published to students. Staff might advertise their projects in lectures and seminars, or be available to discuss on a one-to-one basis. From this list students are expected to select, in rank order, their preferred project; then they have to wait to have their project confirmed, and may well be disappointed if they have selected only popular topics. Each student is allocated a project (hopefully) in their highest ranking preference, and on they go. The advantage of this method is that the allocation can be moderated by various criteria. For example, for an undergraduate cohort, one could prioritize on the selection made by students who are returning from a period on placement. Alternatively, one could allocate projects on a first-come-first-served basis—strict time order of selections received. A further example might be to promote meritocracy and allocate the projects in the order of the students' grade-point-average (GPA) to date. Naturally, some combination of the above is an option.

4.2.6   The Etiquette of Dissertation Supervision

Guide, don't instruct! We discussed this with regard to managing expectations earlier.

Let the student know how often you expect to meet with them—this will be dependent upon the project—and how to arrange their meetings. Once you've made some decision about how it's going to operate, you might formalize the arrangement with each party signing a memorandum of understanding (MOU).

It is possible that your institution mandates that all academic staff supervise some dissertation projects for taught courses. Many colleagues consider this to be a chore, and a distraction from advancing their own research—it takes time in which they could otherwise be writing grant proposals. The sad thing is that it is frequently the professors with the longest publication record who are the most difficult to get engaged with the dissertation project supervision process. This lack of student engagement is highly detrimental to the institution, as the message to the wider world is that professors are insular, only interested in their own research and have no interest in passing on their knowledge and skills to the next generation. We have heard tell, on the other hand, of institutions where all the highly successful professors teach in the first-year of undergraduate courses. They are inspirational. Students attend all their lectures and are wrapped in their attention. End of year results are consistently above passable, and attrition rates are miniscule. Passing knowledge on for professionals involved in vocational courses is a requirement—it has to be done—teaching is not optional. The dissertation project is one means of teaching that can be highly rewarding.

4.2.7   Feedback

A key aspect of supervision is in the provision of feedback to the student. In some cases, the supervisor will be involved in the examination of the dissertation report; in other cases, examination will be carried out entirely independently. Whichever system is in effect, feedback needs to be encouraging, but not hold out false hopes, that is, don't lead the student to believe that they're likely to get an A-grade, if they're not! Similarly, never tell a student directly that they are going to fail, if the work is so bad that it is ungradable. Emphasize that the feedback is about the work, and not personal. Poor progress may come about due to a variety of reasons, not all of which may be to do with the work. As a supervisor, your role is to establish whether the student has what is needed to carry out the project, and anything more personal probably needs you to encourage them to engage with the university's mental health counselling service. If you do have to deliver bad news, it is better couched in terms of disappointment (to themselves, to you, etc.). The sense this induces may be enough to get them working with more diligence. If they respond positively, focus on how work could be improved. If they seem overwhelmed by a sense of failure, then they need the counselling service quickly; this is not your problem to solve.

From an entirely self-interested viewpoint, it is not unknown that students protest dissertation marks, and such protests can include accusations of a lack of professionalism by academic staff. A thorough written or electronic record of ongoing feedback can effectively counter such accusations.

4.3   Advice to Project Clients

4.3.1   The Value of Industry-University Collaboration

There is great value in collaborative projects, both for industry and universities. Some of the advantages of working with students are:

- Potential for innovation—they bring a fresh set of eyes to an existing problem and often challenge the accepted practice
- Development of the next generation of engineers
- Opportunity for consultancy at an affordable price.

The disadvantages should not be ignored. Students have no track-record to which you can refer when selecting them for a collaborative project. So, there is an inherent risk as well as potential reward in collaborating with a university-based project.

4.3.2   Finding the Right Collaborator

Finding the right collaborator for your project in a university can be challenging. We suggest you to look

(a)   for an academic (this is, depending upon university and country likely to be somebody with the job title of lecturer/senior lecturer/reader/professor, or assistant professor/associate professor/professor) with research interests in your project's subject area
(b)   in a location convenient for yourself.

An academic is more likely to be receptive to being involved in your project—which they will, if it aligns to their own research interests. You need to decide where to look, which will depend upon how you are going to conduct your relationship with the project student and their supervisor. If you want to monitor progress with monthly meetings, then having them in a place that is convenient for you to visit, or them to visit you, would be much better than incurring additional expense and carbon emissions (this might come down to your mutual preference for remote working methods).

4.3.2.1   Seeking Researchers in the Field

You might consider thinking your way into the problem like a student and do an initial literature review to find out who is doing the current research in your topic area, and where—at which university—they are based. If you are not reading research papers on a day-to-day basis, Google Scholar is a free search engine of publications that is easy to apply filters of dates and countries as well as key words. There is a tale of an engineer who did just this, only to discover that he was reading a paper co-authored by one of his lecturers from years earlier. He had not realized that his lecturer was one of the leading authorities on a particular technique he needed to apply to his problem at work. Research papers always include contact details for a "corresponding author". In addition to this, most academics have web-pages detailing their research interests; so, you could trawl the professional profiles of staff to view other research and teaching interests before you make contact.

4.3.2.2   Using Your Alumni Connections

If you have been through the engineering education system yourself, you could try your own old university. If you are a fairly recent graduate (or have recent graduates working with you), then dealing with a staff member you know might be a relatively straightforward means to find a collaborator. If you're not in contact with your old lecturers, then you (or they) could try contacting the alumni office, who usually would be delighted to hear from their graduates—although you may have to give an interview to their marketing department in *quid pro quo*.

4.3.2.3 Supporting Local Students

If your priority is actually to support a local student—perhaps with a view to finding a future employee—then seeking a collaboration in your local university might be the way ahead. Our general advice here is don't cold call asking someone to speak about "student projects". Universities—and this is true the world over—are like hydra. There is too much scope to get passed around in an ever-increasing spiral of misdirection.

You could try the university's "research development office" (they go by various names). These offices exist to make research links with industry; usually (but not always) they are concerned with major collaborations, and are not able to deal with one-off dissertation student projects; however, there is scope to get involved in research whatever your size of company. Increasingly, universities are recognizing that small-to-medium-sized enterprises (SMEs) are also wealth-creators, and for many universities, part of the "development office" is set up to work with SMEs, start-ups and people with an idea. It is wise to be candid about the size and scope of your project. If they don't respond positively to your idea immediately, do ask to be referred to the student-projects coordinator in their engineering department.

An alternative way of getting an introduction to potential collaborators in a university is through the careers office; as every institution has concerns about the employability of its students, they are generally very receptive to involvement from industry. Possibly the most effective means of making contact is to offer a guest lecture—ideally, this should not be all about how wonderful your company is, but rather about the challenges and opportunities in your industry. It could legitimately be about the problem at the centre of your potential project. There is a slim chance that someone in the audience will want to pursue your project; however, your objective should be to make contact with an ideal supervisor—an enthusiastic student should be able to make this introduction.

It is not just the careers office who organize guest presentations. This is something student clubs and societies also often do. This can be an alternative way into the university, more likely to engage students. Look out for the specialist societies, such as the Robotics Soc or the club that enters the drone flying competition, or Engineers Without Borders. These are often the places you'll find highly motivated engineering students who can demonstratively organize their time, are willing to apply their engineering know-how to solve problems and not just those within the limits of their degree course

4.3.2.4   The Student in Your Workplace

If your organization supports continuous professional development, then you might have an engineer on your payroll who is studying for a degree, either by distance learning (DL) or part time (PT). Alternatively, it might be that your company has got a registered full-time student, who is completing a period of employment (placement or internship) with you as part of their course. This is a relatively low risk way of recruiting a student to carry out a project for you, as they are already in your employment, so there is already a relationship of trust between your organization and the student. However, if you are not the student's line-manager, you may need to negotiate with such to get cooperation to work on your project.

4.3.3   Managing Expectations

As with any project, it is important to define

- Time
- Scope
- Budget and resources.

4.3.3.1   Time: Timing and Commitment

The academic year is very inflexible. Your project may have immediacy for you, and if you time it right, it will get picked and run with swiftly. This is best talked over with your prospective academic supervisor—as they will have the most experience of fitting in with the academic year's timeframe. There are multiple start times and finish times of degree courses in operation. If you have time constraints on your project, make it clear to the prospective supervisor what they are—this may affect their decision about what type of student (undergraduate or MSc) carries out the project, or whether it is feasible at all.

It is good practice at the start of any project to establish how much time you can commit to the project. Like any project in the workplace, this is a matter of striking a balance—enough to monitor progress, but not too much so that you are micromanaging the endeavour.

Set some milestones. Discuss with the student and their supervisor what the deadline of the project is for the university; what plans they have for any absences during the project programme (factor in absences for university closures, and any local holidays). There are important life-lessons about work–life balance that your student may not have yet learned. You might have your own target dates for key achievements, such as getting design drawings to your manufacturer, or having a test rig assembled. Check that your dates are clearly noted within the schedule, and allow yourself sufficient time at these points to provide feedback to the student on their progress. You might decide not to meet in person, but some scheduled contact is important so that the student does not feel abandoned.

4.3.3.2 Scope of the Project

At the outset of your project, it will greatly help if you can work out a statement of scope. You might already know exactly what is the problem that you want to solve, or you might have a clear aim in mind. It is best to discuss this with your project student and their supervisor, in order to define a project that is feasible AND will meet the learning outcomes for academic purposes.

Be clear about your expectations from the project. Quite often, a project in collaboration between a company and university requires a report that can be handed over to a senior manager, or forms the basis of a report that goes to the board of directors, or to a client. The final report of the dissertation project is not necessarily this report. The dissertation project report might contain all the information you need, but it might contain a lot of information you don't want to share. The nature of the dissertation is that it should contain a balanced argument. Occasionally, in industry this is perceived to be extraneous information.

Similarly, the dissertation report must clearly show, by suitable citation to references, where the existing ideas have come from. References also are needed as evidence to support the context of work. It is unusual to have such detailed literature reviews in business documents. So, if you are looking for a certain type of report for your business, you need to make it clear at the outset. The student's project may well be about creating your business report, but they will also need to create a dissertation report about creating your business report, which will be for the assessment of their degree.

4.3.3.3 Budget and Resource Considerations

Best (learning) practice suggests that you let your project-student work out what resources they need and calculate what budget is required to complete the work. This is usually part of their requirement to develop a project proposal. You might help them along if you have, or have access to, specialist equipment. Support in-kind, rather than financial support, is often more beneficial. It is possible that the student has access to a notional budget for items that relate to their education, such as software licenses, materials and manufacture relating to their project, but your contribution might enable more versatile software, or permit higher quality materials, or permit more sophisticated forms of manufacture.

Note that if your contribution in-kind is permission for the student to use your workshop or laboratory facilities, there may be insurance liability considerations. It may be that the student is covered by the university's insurance whilst they are working on a university project, but this needs to be checked before work starts. Engineering students are usually required to carry out a risk assessment before starting practical work—this will probably coincide with what is needed in your workplace. It is certainly a valuable lesson for them to see what best practice is in reality.

4.3.4   Etiquette

If you have any concerns about the conduct or behaviour of a student on a project, this is probably best addressed in consultation with their project supervisor. For many students, working on a project sponsored by industry is their first encounter with the grown-up workplace and the engineering business world. If they have completed a placement, or worked before this university course, they may have a little more worldly wisdom. You may have forgotten what it was like to be so unformed. An example of this was when a project client reported to a colleague of ours that he was upset by the behaviour of a student who kept looking at her smartphone during their meetings. He thought it was very off-hand that she should be checking her social media whilst he was trying to ascertain what progress she was making on his project. When the supervisor queried her on this point, she was devastated that the client thought she was being rude—she had been consulting her notes, which were on the phone. Different generations work in differing ways—that's why we need to work together!

4.3.4.1   Agreed Communication Modes

At the start of the project agree what your normal mode of communication is. It might be best to have this conversation with the project supervisor. If, for example, you don't want to be bothered by phone calls, then make it clear you would prefer emails (or some other means of communication) that you can respond to in agreed time.

4.3.4.2   Hosting Visits

It's a good idea at the outset of a project to establish whether it is possible for the student (either with or without their supervisor) to visit your site. You might consider your office to be very ordinary, but to a student with little or no experience of employment in engineering, it is an opportunity for a valuable insight into the engineering working environment and will make a big impression on them. Don't worry if it can't be managed!

4.3.4.3 Dealing with Commercial Sensitivities

If your project is classified as "commercially confidential", then this should be raised
with the project supervisor as soon as possible. Universities are frequently involved
with collaborations that require formal documentation and agreements. If the project
is not highly sensitive, you might consider using a Memorandum of Understanding
as a means of recording your agreed intentions. If confidentiality is needed, and you
propose use of a non-disclosure agreement (NDA), it can be problematic to identify
all the people who need to be included in a NDA, for example, supervisor, student
and technical support staff, and to obtain university-level agreement. If at all possible,
manage the sensitive information on your side of the collaboration, and avoid putting
burden on the student.

4.3.5 Checklist

It is suggested that the client and student collaborate to complete this checklist at the
start of the project.
About the project:
Please complete one or other of the following statements:

a) The aim of this project is…	b) The research question for this project is…

Challenge yourself to limit the description to 280 characters!
The project can be categorized as ………………………………
Refer to Sect. 4.1.2 for descriptions of the categories of student projects.
About the client:

Personal name:	Formal address:	
	Preferred appellation:	
Organization name: (if relevant)		
Job title:		

Contact details:

Phone no.:	
Alternative phone:	
Email:	
Skype ID:	
Other:	
Postal address:	

If any work is to be carried out away from university, then record Location details:

Name of building or site:	
Street address	
City	
County	
Country	
Code	

If work is away from the university, then a risk assessment is to be submitted to:

Name of person:	
Contact phone:	
Email:	

This should also be copied to the university supervisor /technical manager /chief technician.

Chapter 5
The Opening Literature Review

Abstract All engineering projects must commence by developing a good under-standing of the state of knowledge of the topic: this is known as a literature review. A dissertation always requires an opening formal literature review, likely to require information from a wide variety of sources including textbooks, academic journals, lecture notes and engineering standards. This must be formally documented, with each source referenced, described and weighed for its qualities. Subsequently this literature review will require further updating and become a major opening section of the report.

© Springer Nature Switzerland AG 2020   59
P. Gratton and G. Gratton, *Achieving Success with the Engineering Dissertation*,
https://doi.org/10.1007/978-3-030-33192-4_5

5.1 Beginning the Quest

As you start your project, it's certain that you won't know everything you need to know about the topic, or about how to solve the problems in it. It's also certain that the answers you need won't be all in one place—so even if your supervisor directs you to a textbook or some papers to read (as they often will), it'll only be a small part of the story.

Hence there's a need for a *literature review*, where you read, digest and record the opening knowledge needed to proceed with the project. That literature is likely to fall into four categories:

(1) Textbooks, encyclopaedias and course notes
(2) Academic literature
(3) Engineering standards
(4) "Other" sources, such as magazine articles, websites, movies, survey results or newspaper reports.

Another sub-division that you'll often hear and refer to is "primary" versus "secondary" sources. This matters a lot in humanities, perhaps less so with engineering as—unlike historical events or the work of a great artist—engineering labwork and analysis can be repeated. Nonetheless, some information (such as observations of an accident, incredibly expensive results from the International Space Station, or CERN,[1] or survey results from users) will still be regarded as "primary" in the classical sense: on the other hand, all correct proofs of Bernoulli's equation or explanations of Young's modulus are, arguably, equal and "secondary".

A lot of the information you're searching for will be on web resources (either in the general World Wide Web, or closed resources such as your university library or the members area of your learned society's website), but it's important as we go down this journey to remember two very important things.

#1 Not everything you need is on the internet!

#2 Not everything you find on the internet is a website—sometimes it is a paper, report, or other formal document that simply is made available through the internet.

The difference on this second point is that documents that *happen to be* on the internet may exist in other places, and unlike websites should have a fixed amendment state. Please keep these points in mind as you progress with your literature review.

To put the literature review together you need a starting point. The best person to advise you on that will be your supervisor (Fig. 5.1) who is likely to have a good understanding of where a good starting place will be. Very commonly this will be in academic literature: journal papers or refereed conference papers. Very helpfully, most academic literature—to a much greater extent than most textbooks and course notes will contain long lists of references. Those references provide an excellent trail that can be followed to start to build a list of relevant literature.

---

[1]CERN—*Conseil Européen pour la recherche nucléaire*, or European Nuclear Research Centre—in Switzerland: home of the famous Large Hadron Collider.

Fig. 5.1 Asking the
supervisor for assistance
starting the literature review
(Mr William Heath
Robinson, My Line of Life
(1938), p. 35)

5.2   Searching and Making Notes

The short-term objective will be to know much more about the topic of your dissertation, and thus be far better set up to progress the engineering or research work that
will be the core of your dissertation project. This will be demonstrated by a written
document that is likely to be assessed by your supervisor (whether it actually attracts
a mark will depend upon your university's assessment procedures; it's sensible to
act as if it does. But, in all cases it's the academic feedback that matters the most:
the objective is to <u>learn</u>)!

   You will need to do a lot of reading—although resist the temptation to read
everything you find from cover to cover: whilst tempting, you simply won't have the
time to do that with all of your source documents. What you must do from the start
is construct a document that consists of notes for each reference on what the most
critical information was for your interests, the *context* of that information and (very
importantly) the information you'll eventually need if it's to be included in your list
of references that where you found the document—so that you are able to go back
to it later if you need to.

   As you work through the documents, you should start very early to write a summary of what each tells you <u>about the topic of your dissertation</u>. Early on this will

be fairly scrappy, and it doesn't really matter if you jump around in format and layout: that can be sorted out later. Nor does it matter whether you work in a word-processor, longhand or in some other application (like RefWorks)—this detail can be sorted out later. What is vital, however, as you go along is that you assess and record the following for each source of information:

(1) What does the reference tell you about the topic you are researching?
(2) What's the context in which it says this? For example, is this information in a sales brochure, an engineering standard, a journal article or a magazine article—each of these clearly need to be regarded and treated differently. (This assessment of context is often referred to as *critical analysis*—not to be confused with *criticism*—your task is to assess, like a film critic, not simply be negative.)
(3) Where did you find the information source? [Both the formal reference and how you obtained it yourself should be noted; for example, a location in the library, or a URL.] Where it's something relatively long like a book, also note where within the reference (Chapter? Page number?).

You'll be following multiple trails of logic, including the references in journal and conference papers, standards mentioned in textbooks—and pieces of narrative that take you to research new sub-topics. This is fine—serendipity is inevitable; you will find eventually that you'll end up not using quite a lot of the work you've put into this, but that is also inevitable and not something to be upset about when it happens. The only rule here is not to get too carried away with things that are *only* interesting, rather than relevant—as time spent doing that you won't get back. If in doubt, just make a note of possible interesting avenues to pursue later, just for fun, when you have more time available.

At the start, your literature review will probably somewhat resemble a shambolic "brain dump"—that's okay, so long as it doesn't stay that way indefinitely. Usually it's better to have all the information down, then sort it out—than force a rigid structure before you know what all your information looks like.

5.3 Structuring the Literature Review

The literature review should, *eventually*, be well structured with meaningful subject headings. Unless insisted upon by your university's guidelines (which of course always take precedence over anything in this book), try to avoid ultra-simplistic headings like "Introduction", "Body", "Method". Instead, try and use more meaningful headings such as "Why we need hybrid-engine washing machines", "Existing research in hybrid powered white goods" and "The scope for development of a new coal/steam dishwasher". Your reader can much better understand what you're really talking about with this level of detail. Of course, these headings should be well matched to the stated objectives of the project—if you can't match them up, you may be writing about the wrong thing.

5.4   Composing the Literature Review

The literature review for a project looking into some aspect of climate change might start like this:

Why research climate change impacts on aeroplane performance

Not only do aviation emissions affect climate (IPCC 2014), aviation is also affected by changing atmospheric characteristics. Mean tropospheric temperatures are increasing, and mean lower stratospheric temperatures are decreasing (IPCC 2014).

Jetstreams and stationary wave pattern winds at cruising altitudes are changing, particularly with increased carbon dioxide induced radiative forcing. Modified wind shears consequently modify both mean strength and frequency of clear-air turbulence (Williams and Joshi 2013; Williams 2017). This will also modify journey times (Karnauskas et al. 2016; Irvine et al. 2016).

References

Intergovernmental Panel on Climate Change 2014: Climate Change 2014: Synthesis Report. Contribution of Working Groups I, II and III to the Fifth Assessment Report of the Intergovernmental Panel on Climate Change

Irvine E.A., Shine K.P. and Stringer M. A. (2016), *What are the implications of climate change for trans-Atlantic aircraft routing and flight time?* Transportation Research Part D: Transport and Environment, Volume 47, 44–53. https://doi.org/10.1016/j.trd.2016.04.014, 2016

Karnauskas K.B., Donnelly J.P., Barkley H.C. and Martin J.E., Coupling between air travel and climate. Nat. Clim. Change 5 1068–73, 2016

Williams P.D. and Joshi M.M., Intensification of winter transatlantic aviation turbulence in response to climate change, Nat. Clim. Change 3 644–8. https://doi.org/10.1038/nclimate1866, 2013

Williams, P. D., Increased Light, Moderate, and Severe Clear-Air Turbulence in Response to Climate Change Adv. Atmos. Sci., 34(5), pp 576–586, 2017. https://doi.org/10.1007/s00376-017-6268-2

The method here for citing references from the text (which in reality would be either at the end of the report, or the section) is called *Harvard*, named after the university and is one of several common methods—perhaps the most popular for academic work. You'll note that the references are a similar length to the text itself—that is not a problem if it occurs—but neither is it a target. Describing what references say, and explaining your *critical* views on them at greater length is good.

Note how the text reads fluently, explaining and discussing the topic. References are called out (or cited) within the text at the relevant points where the information is used.

When using Harvard, it's normal to list the references at the end of the dissertation in alphabetical order of author(s) names, then by date if more than one reference of the same author. Other systems may use a numeric system, in which case it's normal to list them still at the end, but in the order they've been referenced—most word-processors have tools to make this very easy. There are also apps available that can store references and import them into a document in whatever format you want them—check what your university has available, although these aren't mandatory (and for the record, neither author of this book uses one of those apps, preferring to use what's native to the word-processor).

Note that in this book we put the references in each chapter, not at the end of the book. This is to help you, particularly if downloading a single chapter of the book to work with—it's most likely not the right way for you to write your own reports, which will usually require all of the references at the end of the document.

Your university's guidelines should tell you how long to make your literature review—but lacking any specific guidance, around 1000–2000 words with 20–40 references is likely to be in the right order, probably a bit more for postgraduate dissertations.

5.5 Formally Recording a Reference

There are multiple systems of recording a reference, but they all come down to the following information:

(1) Who wrote it? (if there's a single author—and usually up to three authors, list all of them; if the list gets longer than its normal, just give the first author then "et al." which is a Latin phrase for "and everybody else".) Remember that whilst all documents have an author, sometimes the author may be an organization, not a person.
(2) The title (usually the one thing that doesn't get abbreviated).
(3) A reference for the document. If this is a journal that's likely to be the volume and paper or page number. Depending upon referencing system, for books this may be a publisher and city, or may be an ISBN (International Standard Book Number).
(4) A publication date, or where information was on a website—the date it was accessed.

As a matter of both good practice and ticking of academic boxes, your references will almost certainly want to be a reasonably balanced mixture of all four of the categories we introduced at the start of this section. If everything is from a single type of source (particularly web sources), then you may be going wrong. At the same time (and we've said this before), always remember that just because something was found online, does not make it a website. So only use URLs as your reference if something genuinely is a website. This, of course, does not mean that you shouldn't record a URL in your own working notes, just not usually in any formal document that you are handing in.

We've provided examples of how to reference various types of documents in Appendix A at the end of the book. There are several common referencing systems, and your university may require you to use one particular system—but the universal rule is: be consistent. So long as you always use the same comprehensive system of citing and referencing, then that aspect of your writing is likely to be well treated by readers.

5.6 Maintaining the Literature Review as You Work

The point where you hand-in your literature review to your supervisor is not the end of it, just an important landmark. Their feedback will be important, both because they will have the experience to show you how to improve it (or learn from where you've done a good job), as well as be able to comment on whether you've taken the right directions with your thinking. They should also, with their experience, be able to advise whether the direction the literature review indicates your work is going is appropriate—and whether you appear to be embracing too little, too great or about the right amount of scope. (The most common fault, incidentally, is to try and cover too much scope—the relatively limited number of hours you can afford to spend on your dissertation means that scope must be quite constrained, and the literature review provides a great opportunity to trap over-ambition before it becomes a problem).

So, almost immediately there's been feedback, you'll probably want to make some immediate improvements, to trap the main part of the feedback. If the feedback was very critical or came with poor marks, then that might be more work—but don't get despondent if that's the case. Improvements made early on are cheap, in terms of time commitment, than solving a major redirection later; even if the marks count towards your final grade, they're probably not a large proportion of it. You may also be told that a significant part of the literature review isn't necessary and wants deleting—again, don't get upset by this, but make sure as you make amendments you keep all the old versions (by this point, almost certainly electronically—so it's very easy to do).

Reminder—Always Keep at Least Two Backups in Two Different Places of all Your Computer Files!
As you progress through the rest of the project, you will, however, keep finding new avenues you need to pursue, in terms of reading and understanding all forms of literature. So, the literature review itself will need to develop—both with information being added, and sometimes deleted as you decide information is no longer relevant. Eventually, it will be a major part of the dissertation report—most likely either within or immediately following the introduction.

Chapter 6
The Project Plan

Abstract All projects require planning for time, resources and finance: this is equally true for a dissertation. The project plan should use a numerical network analysis method from which a Gantt chart may be created. All resources must be identified, and their availability and access should be planned ahead. All financial costs must be estimated, approved, then any purchasing put through channels approved by the university; subsequently actual expenditure must have been calculated and summarized.

© Springer Nature Switzerland AG 2020   67
P. Gratton and G. Gratton, *Achieving Success with the Engineering Dissertation*,
https://doi.org/10.1007/978-3-030-33192-4_6

6.1 The Project Plan and Time-Management

6.1.1 Introducing the Project Plan and Planning

The first thing to understand about your project is what it is about and what work needs to be done. This is the scope of the project.

At the end of the project, you will have to hand-in a project report (often referred to as the dissertation or "disso"), the assessment of which will contribute the major portion of your final mark; however, you should note that the report is NOT the project! Writing up the report is only part of the work that needs to be done. What you need to establish now is what you are going to be doing in order to have something to write up!

The project starts as soon as you start thinking about it, and talking about it. If you have selected a project that has been proposed by an academic member of staff, then the likelihood is that their title was accompanied by a very brief description about the project. This will hopefully give you an insight into the nature of the project; for example, it might indicate whether it involves computer modelling, designing and building something, or getting into the lab to take measurements.

Every engineering project must have some form of investigation and analysis of an engineering problem.

A good place to start your project is by talking with your supervisor. Talk about what your motivation is for choosing this project. Check with them that you understand what the project is about, and what you think you are going to be doing.

6.1.2 Preparing the Project Plan

Your supervisor might ask you to do some background reading, or provide you with specific references to go and look up. It is often the case that your project is a continuation of a piece of research that was started a while ago. It might be that the project you are to work on was recommended by a previous project student. However, do not be surprised if your supervisor is unwilling to share a report from a previous student. The reasons for this vary—perhaps, it might be a poorly written report or your supervisor doesn't want you to be confused by it; or perhaps they don't want you to use the report as a template for your project.

In the early stages of project, be prepared to ask lots of questions of your supervisor. They won't mind. In fact, a new project student who is not asking questions is a worry for a supervisor—is the student interested in the project? Do they understand what the project is about? Are they going to do anything useful?

First steps on your project are simple:

1. Read
2. Write
3. Repeat

Read—Before you first meet with your supervisor, it's a good idea to feed some of the keywords relevant to your project into your favourite search engine. Find and read good introductory material.

Write—As supervisors, having met with a project student for the first time, we usually ask them to write a 200-word summary: telling us what the project is about, explaining how they are going to tackle it and describing what they are going to, need to do or use (e.g. software, lab space, site visit, etc.). We ask for this to test our mutual understanding of the project topic, whether the student can summarize ideas coherently and comprehensibly, with no errors of spelling, grammar and punctuation. This gives us an insight into their general attention to detail.

Repeat—Seek out more technical, relevant publications using more specialist search tools; use the university library; learn qualities of the literature.

6.1.3   A Proposal

Composing a project proposal may be the first task your project supervisor is likely to ask you to complete, or it might be a requirement of your course and forms part of the assessment. Your course may have its own requirement for the project proposal, but they will generally be along the lines of containing the following:

- Aim
- Objectives
- Context
- Methods
- Project Management

    - Time schedule for tasks
    - Risk assessment and ethical approval
    - Resources (lab, equipment, technician)

- References

Ideally, your project should have a clearly declared simple aim. For undergraduate projects, that is, involving students who have never undertaken a significant project on their own, it is recommended that supervisors limit them to ONE aim with maybe three subsidiary objectives. For postgraduate students, who are building upon their previous experience as undergrads, then they might be encouraged to stretch their zone of experience (and comfort), and take on a project with multiple aims (such as design-make-and-evaluate). Generally, it is recommended to keep the focus of your project narrow, and very close to the title proposed. Far too often, a student would like to propose a project on a popular (and extensive) topic, such as renewable energy technologies, and want to consider all technologies, but what they write about demonstrates very little understanding of the implications of the various techniques they are proposing to consider, and have no idea how long it will take to create such a wide-ranging topic. The mantra adopted is that of aircraft designer Kelly Johnson: "Keep it Simple!" (usually referred to as the "K.I.S.S. principle").

6.1.4 Defining the Objectives

At the outset of your project, you may not yet fully understand what the objectives of your project are. Supervisors may often define the aim of the project in a project title, but may deliberately leave off publishing objectives for the project, as a student is more likely to buy-into a project, and succeed at it, if they define their own. For undergraduate projects, it is suggested that the number of objectives are limited to three; Masters-level students are expected to cope well with more complexity, so might be encouraged to plan for more objectives.

Don't get too hung-up on not having objectives in the early days of the project. Start reading around the topic to establish in your own mind what the project is about.

- Read widely to understand why this project is relevant
- Read books, newspapers, journals, trade magazines, blogs, government reports, guidance notes, everything
- Keep notes (including reference details) of what you have read
- Some of what you read will be irrelevant in the end; so, be prepared to delete it!

All this reading develops your understanding of the context of your project, and ultimately contributes to the background or literature review.

Having established the context of the project, it is time to review the aim. With the aim in mind, and an understanding of what has previously been published on the topic of the proposed dissertation, it should be possible to define the research question or specific aim for your project.

6.1.5 Detailed Time Planning

Be aware of the key dates in your project—the milestones—and how many hours are available to you to work on the project each week. There is a time-management exercise later in this section to help you allocate hours to your project.

Even before you have anything to do on your project, make sure you are reading about your topic in the hours you have allocated to the project.

It might help you to allocate dates and times to meet with your supervisor at the start of the project. Some supervisors (or students!) may not want to be committed to these appointments up-front, but there is nothing to stop you putting the meetings in your diary, and asking your supervisor for an appointment closer to the time. Individual supervisors may have their own preference for how frequently they meet with their project students. Depending on the project, meeting about once every 3–4 weeks is enough time for students to make progress, but frequent enough that the supervisor can monitor that they are keeping on track. It is best practice to ask students to write a very brief report of progress (a kind of log), to which supervisors can add suggestions for targets in the next 3–4 weeks. On both signing the report to acknowledge the progress, an effective log of the project is created. Some universities require and assess this document.

6.1.6   Pre-proposal Time-Management Exercise

Here is an exercise for you that might be best carried out using a spreadsheet. Address the following questions:

How many hours are there in a week?
How many hours per week are you doing the following:

- asleep?
- eating, cooking, spending time with family and friends?
- exercising, participating in sports?
- in lectures?
- working on assignments?
- working on your project?
- working for payment?

We asked a group of third-year undergraduate students, and these are the averages of their responses:

How many hours are there in a week? Indisputably, $24 \times 7 = 168$
How many hours per week are you

1.	asleep?	Per day, about $8 \times 7 =$	56
2.	eating, cooking, spending time with family and friends?		21
3.	exercising, participating in sports?		8
4.	in lectures?		17
5.	working on assignments?		34
6.	working on your project?		17
7.	working for payment?		15

6.1.7 How Much Time Should You Spend on Your Dissertation Project?

Each university, or each engineering department, will have their own recommendation for this. The hours allocated to coursework are linked to academic credit—a notion of how much learning is contained in the exercise. Another variable is when deadlines are relative to starting the project, and what else you need to fit in for your course of studies at the same time. Unlike lectures and time spent in the labs, it is unlikely that your time on the project will be on your timetable, so it will be up to you how you manage it. Demonstrating that you can manage your time is one of the learning outcomes of the project, and it can in itself be a big challenge.

The example represented by the graph in Fig. 6.1 is a suggestion for a project that is supposed to be completed in 400 hour stretched across an academic year that is divided into two terms of 12 weeks, and has a three-week break for holidays in the middle. For this example, it's assumed that the study-week has 40 hours of full-time study. The usual full-time degree course consists of 1,200 hours per year, but only a small percentage of this will be scheduled on your timetable as "contact-hours".

This example project is "worth" a third of the credits for the academic year—a non-trivial proportion. It is suggested that you might "front load" the time spent on the project. The early weeks are spent reading, planning and organizing. It is also suggested that you plan sometime away from the project in the holiday season, and return to it refreshed. Of course, you might need to check on the progress of purchase orders in this time, so an hour a week is shown in this example. The chart in Fig. 6.1 helps you to visualize how you are going to allocate your time. In industry

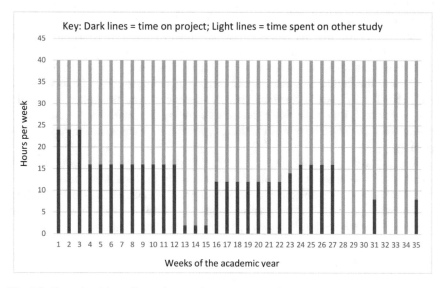

Fig. 6.1 Example of time allocated to a project over one academic year

you will usually be required to fill-in a timesheet for each project and non-income generating activity you are engaged in. You might set up a similar monitoring system for yourself. It is certainly best practice to keep a log of progress, which will form the basis of reflection in future evaluation of the project. If your weekly hours on the project exceed the expected average, perhaps you need to consider what other coursework is suffering.

In this particular example, the final report is handed-in in Week 27. This is followed by an oral presentation in Week 31 and a poster competition in Week 35. It is sometimes questioned whether these assessments are actually part of the learning. In our experience, they are most definitely learning experiences, and so the time spent on them should be included in the project's time-budget.

6.1.8 Using Project Management Techniques and Setting Milestones for Your Project Plan

As an engineering student, we hope that you've studied project management techniques during your degree course, So, we're not going to dwell on the merits and demerits of systems such as PMP[1] and PRINCE2.[2] You could use a software package, like Microsoft Project, but even this is designed for relatively large projects and tends to take a great deal of time to set up. Hence, we suggest a return to first principles and use a version of the critical path method (CPM).

The time-plan needs to do several things: set out the sequence of tasks, identify which tasks are contingent upon others and identify which tasks are on the critical path. Over the years, many textbooks and guidance websites have been created to explain how to successfully create a time-plan, so we are not going to reiterate the whole process in detail here. This is a quick-and-dirty routine. The guru of time-plans was an American mechanical engineer named Henry Laurence Gantt (1861–1919)— his own writing about the science of project management provides an historical insight into the building of America through major infrastructure projects in the 1910s. His legacy to us is the Gantt chart. The result of your time-planning should be a Gantt chart, which should be treated as a "living document" and updated on a regular basis. It should be one of the things you refer to when you have supervision meetings.

You can't really plan your time effectively until you know what tasks you are going to carry out in order to meet the objectives of your project. So, it is a priority in the early week(s) to decide objectives and define the tasks (usually tasks can be bundled up into "work packages", sometimes referred to as "work blocks"). It will help you plan your time tremendously if you can identify the "milestones" of your project.

---

[1]PMP stands for *Project Management Professional*, a standard set by the Project Management Institute, an international organization based in Newtown, Pennsylvania.

[2]PRINCE2 stands for *PRojects In Controlled Environments*, a method and practitioner accreditation scheme initiated by UK government for infrastructure projects, and adopted in many other countries.

Milestones are points in the project when key tasks occur, or have been completed. They give you an indication of how close you are to completing the metaphorical journey of your project to completion. Exactly what these are and when they occur in your plan will be specific to your particular project; however, make sure that you are aware of all the milestones that are imposed upon you by your department, and, where appropriate, your project client.

Each engineering department is likely to have a different way of working, but it is often the case that an interim or progress report will be required—it might even form part of the assessment. It is usual in industry on any project lasting more than a couple of weeks that you meet with the "owner" of your project to review progress—these progress reports can be used to set some milestones—so that in scheduling tasks, you aim to have some much completed in time for a particular report. Other milestones might be defined by things like completion of an assembly, switching on a machine for the first time and taking delivery of a key piece of equipment. All these milestone events will be on the critical path—that is, they MUST happen in order for the project aim to be achieved.

6.1.9   Using CPM on an Individual Project—One Participant

All the reading around the background of the project that you do in the early weeks will help you plan the details of what work you are going to carry out. Once you have an idea of the work, you need to divide it into a list of tasks. This list can then be used to create your time-plan.

By way of illustration, consider a project codenamed "Inlet Route Weighting" (IRW), which has 19 distinct work packages. These are, labelled WP01—WP19 and their details are listed in Table 6.1. For simplicity, some milestones have been included as activities. These are working backwards: the final report hand-in (WP19) and the interim report hand-in (WP12). These are scheduled for 30th March 2020 and 13th January 2020, respectively. These are independent, fixed points in the project's timeline, and will incur a penalty if missed. There is one other milestone in the project, but this is dependent upon other work packages being completed, and this is the receipt of equipment (shown as WP07). In actual fact, the activity of WP07 is an estimate of the lead-time waiting for the order to be fulfilled and delivered. This event is critical as none of the tests in the project can be run until the equipment is assembled into the test rig.

Another way of visualizing the project is to arrange the activities into a network. Figure 6.2 is an attempt at this—although it has to be modified to get it into a page. As you sketch out your activities into a network, it may become obvious (as in this case) that most tasks are dependent upon one prerequisite task, so you get a sort of "string of beads" network. It also becomes apparent that there is no easy way to deal with the interim report (WP12), which isn't really dependent upon any task, only completion by a hand-in date. Another quandary is dealing with the lab closure from 16th December to 6th January; fixed dates and out of your control. In the

Table 6.1 Example of activity details for an individual project

Item	Description	Duration (weeks)	Contingent upon
WP01	Reading	3	Nil
WP02	Write up literature review	2	WP01[a]
WP03	Write up methodology	2	WP02
WP04	Design test rig	2	WP03
WP05	Get quotes	0.5	WP04
WP06	Complete purchase orders	0.5	WP05
WP07	Receive equipment	2	WP06
WP08	Assemble test rig	1	WP07
WP09	Carry out pilot study	1	WP08
WP10	Write up results for pilot	1	WP09
WP11	Analyse pilot study	1	WP09
WP12	Compose interim report[b]	1	Nil
WP13	Adjust rig & run tests	3	WP11
WP14	Write up results of tests	1	WP13
WP15	Analyse test runs	2	WP14
WP16	Write up discussion	2	WP15
WP17	Write up conclusions	2	WP16
WP18	Proof read and revise	1	WP17
WP19	Hand-in	1	WP18

Notes

[a]WP02 is contingent upon the start of WP01, assuming writing up the literature review can start almost as soon as reading is in progress

[b]WP12 has a deadline of 13/01/2020, but as it is a report on progress, it is not contingent on any other work being completed. There is a preference to start it as late as possible so that more progress can be reported upon

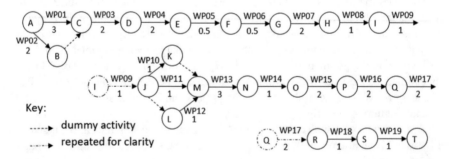

Key:

- - - - ► dummy activity

- - - - ► repeated for clarity

Fig. 6.2 An attempt at creating a network for Project IRW (using the activity-on-arrow method)

network below, node J coincides with the lab closing and node M coincides with the hand-in date of 13th January 2020. Perhaps the lab closure should appear in the network as a separate activity, what do you think? How would you improve this sketch?

Having the durations of tasks and their contingency enables you to calculate the earliest start time, earliest finish time, latest start time and latest finish time for each activity, and consequently to calculate the minimum time that needs to be scheduled to complete the project (a more detailed explanation of this calculation is given later, in Sect. 6.1.11, where a group project having parallel activities is discussed). Hopefully, the end date for the minimum time is before the deadline. Following this exercise you can review the project programme from the perspective of the deadline for deliverables—the end of the project—and work backwards. The problem with the example project in Table 6.2 is that there is no spare time. The duration of tasks and their dependence on each other means that they run right up to the deadline.

Knowing that term starts on 23rd September 2019, and that laboratories are closed from 16th December to 6th January, everything is available to plot the project on a Gantt chart (Fig. 6.3).

Although this picture of the Gantt chart is reproduced at a small scale, you can see immediately that practically everything is on the critical path. This is partly because of the nature of the project and partly because there is only one person carrying out the work. The only item not on the critical path is the interim report, as this can be written during the annual holiday lab closure.

6.1.10 Using Project Run-Offs to Finesse Your (Individual) Project Plan

Ideally, whilst you are still at planning stage, you should review your plan to allow a bit of slack time. Things are very tight in the example project in Fig. 6.3, but there are a couple of issues to be considered before you despair! The first is the timescale you are using; the second is deliberate "project run-offs".

The unit of time for the timescale used in this example is weeks. This gives you a fair idea what you wish to achieve in each week; however, it is not a true presentation of the time to be spent on the work packages. You are targeting to work an average of 15–17 h per week , let's say 16 h per week has been committed to the project, so "Reading" is allocated 48 h, and "Writing the literature review" 32 hours in the same three weeks (80 hours). The reality is that you might comfortably complete both the reading and writing in 72 hours. Later two weeks is allocated to designing the test rig—an estimated 32 hours. This might only take two afternoons in reality—so about 8 hours. Even if these afternoon design-sessions were a week apart, it might be that the process of getting quotations could actually start in the same week as doing the design. At this point, you are fine-tuning your schedule.

Table 6.2 Applying calendar dates to an example project

Item	Description	Duration	Earliest start	Earliest finish	Latest start	Latest finish	Contingent upon
WP01	Reading	3	23/10/2019	14/10/2019	23/10/2019	14/10/2019	Nil
WP02	Write up literature review	2	30/09/2019	14/10/2019	30/09/2019	14/10/2019	WP01
WP03	Write up methodology	2	14/10/2019	28/10/2019	14/10/2019	28/10/2019	WP02
WP04	Design test rig	2	28/10/2019	11/11/2019	28/10/2019	11/11/2019	WP03
WP05	Get quotes	0.5	11/11/2019	18/11/2019	11/11/2019	18/11/2019	WP04
WP06	Complete purchase orders	0.5	11/11/2019	18/11/2019	11/11/2019	18/11/2019	WP05
WP07	Receive equipment	2	18/11/2019	02/12/2019	18/11/2019	02/12/2019	WP06
WP08	Assemble test rig	1	02/12/2019	09/12/2019	02/12/2019	09/12/2019	WP07
WP09	Carry out pilot study	1	09/12/2019	16/12/2019	09/12/2019	16/12/2019	WP08
WP10	Write up results for pilot	1	16/12/2019	23/12/2019	16/12/2019	23/12/2019	WP09
WP11	Analyse pilot study	1	16/12/2019	23/12/2019	16/12/2019	23/12/2019	WP09
WP12	Compose interim report	1	23/12/2019	30/12/2019	06/01/2020	13/01/2020	Nil
WP13	Adjust rig & run tests	3	06/12/2019	27/01/2020	06/12/2019	27/01/2020	WP11
WP14	Write up results of tests	1	27/01/2020	03/02/2020	27/01/2020	03/02/2020	WP13
WP15	Analyse test runs	2	03/02/2020	17/02/2020	03/02/2020	17/02/2020	WP14
WP16	Write up discussion	2	17/02/2020	02/03/3030	17/02/2020	02/03/3030	WP15
WP17	Write up conclusions	2	02/03/2020	16/03/2020	02/03/2020	16/03/2020	WP16
WP18	Proof read and revise	1	16/03/2020	23/03/2020	16/03/2020	23/03/2020	WP17
WP19	Hand-in	1	23/03/2020	30/03/2020	23/03/2020	30/03/2020	WP18

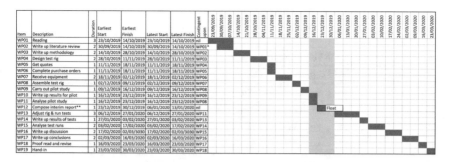

Fig. 6.3 Typical arrangement for a Gantt chart

The other aspect you should programme-in to your plan is run-offs. These are aspects that could be deleted from the plan without appreciably harming the overall aim of the project. A consequence of invoking a run-off is likely to be that you need to re-write the objectives of the project (because time will not permit all of them to be achieved as anticipated at the start of the project), or the method statement for various tasks. There is no harm to the final project report in doing this. This action permits you the opportunity to comment in the discussion chapter that you would have performed the deselected work package had time permitted, and you can recommend it as further work.

In the example project shown in Fig. 6.3, there might be scope for identifying run-off in WP13 "Adjust rig and run tests". Three weeks of time have been allocated for this. It would be worth considering how many times it is proposed to run the test, and how long each test run takes. The question to be addressed is whether reducing either, or both, the number of runs or the duration of each run would significantly compromise the validity of the set of results.

6.1.11 Using CPM on a Group Project—Team of People

The advantage of scheduling a group project is that you can run some activities in parallel because you have more than one person carrying out work. The disadvantage is that you need to plan work so that it runs smoothly without bottlenecks occurring (where other team members are waiting on one person to complete a particularly crucial task). The reality is likely to be that you actually operate your group project as a collection of sub-projects that interlock at clearly identifiable points. The same principles of project management apply for group projects that were discussed earlier for an individual project. In the same way as before, you need to start the process of planning time with a list of activities, which you can develop by adding in their prerequisites and durations. From this information, you can start to sketch out a network of activities, and from this calculate the critical path for the project.

Table 6.3 Example of an activity list for a group project

Work package code	Activity	Pre-requisite	Duration (days)
A	Form research question	–	4
B	Literature search	A	7
C	Select survey points	A	10
D	Draft questionnaire	B	8
E	Train in focus-group leading	B, C	12
F	Arrange focus groups	C	7
G	Conduct survey	D, E, F	5
H	Analyse data	G	4
I	Add co-worker's papers	G	2
J	Present paper to supervisor	H, I	1

Let's consider an example that is a sub-project within a much larger design-make-and-evaluate project using the activity-on-arrow method. The DME project (codename Wheelyfast) is about the development of sports wheelchairs for Paralympic events. The sub-project is a survey of user experience in order to inform future design. A team of six students are carrying out the project. They have identified a particular athletic association, who are keen to work with the team, and have gained ethical approval from the university. The next step is to create an activity list (shown in Table 6.3):

The next step is to represent the activities in the form of a network.

Representing activities in a network
Each activity is represented by an arrow, starting from a point on a small circle (a node) and finishing as an arrowhead. In general, the length of the arrow shaft is not constrained by any time scale. It may, therefore, be varied so that the arrowhead of a preceding activity may be terminated at the circle marking the start of the succeeding activity. In this way a sketch network can be constructed by trial and error. Arrows should ideally not cross one another, and some forethought is required to avoid this.

Numerous computing packages are available to assist in network analysis, and it is possible to key the data into one of these, or to process it numerically using prepared worksheets. However, there is merit in actually sketching out a network by hand, as this helps to check the logic and may reveal errors and omissions in the activity list. Figure 6.4 shows the relatively simple network obtained by applying the activity-on-arrow method to model the UX survey for Wheelyfast.

Calculate the early time and late time of each node
The empty boxes have been added to the network diagram ready to record the early time (ET) and late time (LT) of each event (or node), which represent the start and finish times of activities. Dummy activities (ones that have zero duration) are added to maintain the precedence relationships (Fig. 6.5) of activities (e.g. leading up to

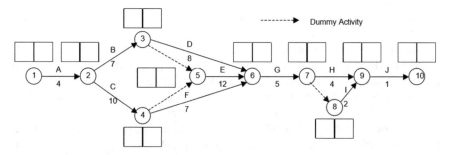

Fig. 6.4 Activity-on-arrow network for UX sub-project for Wheelyfast

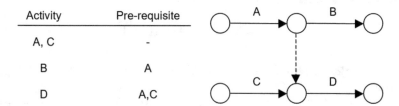

Fig. 6.5 An illustration of the use of a dummy activity to maintain precedence

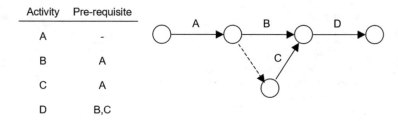

Fig. 6.6 An illustration of the use of a dummy activity to avoid ambiguity

activity E), and where needed to avoid ambiguity (Fig. 6.6).

The early time (ET) of an event is the earliest time by which all activities leading up to that event can be completed. Starting with zero for the first event, you work your way through the network calculating the early time for each activity.

The late time (LT) of an event is the latest time at which activities leading from an event can start without delaying the project. To calculate these you have to work backwards through the network, having set the late time for the final event equal to its early time.

An activity is said to have "slack" when the LT is greater than the ET; it is the amount by which an event can be delayed without automatically delaying the completion of the project overall. Those events for which the slack is zero are said to be *critical* with respect to the overall project duration, and to lie on the *critical path* through the network.

All other events must lie off the critical path, and can be delayed by an amount of time equal to the total slack, without automatically delaying the overall project duration. The occurrence of slack time associated with such *non-critical events* gives rise to a property termed *float* which is associated with the corresponding activities, as they can "float about" in the slack time that is available.

Note that when calculating you might expect the critical path to pass through all those events for which the ET and LT values are equal. This is not so. In the case of the UX sub-project for Wheelyfast, there are two routes: one via Activity F, and the other via Activity E, which satisfy the condition of zero slack above. It is obvious from Fig. 6.4, however, that the critical path is

$$A \rightarrow C \rightarrow E \rightarrow G \rightarrow H \rightarrow J$$

Calculating the Earliest Start Times (EST) and Earliest Finish Times (EFT) for activities
There are five rules to be applied in order to calculate the EST and EFT for each activity. They are as follows:

Five Rules for Earliest Times

Rule 1:	Start at the top of the table and complete the first two columns (EST, EFT) together
Rule 2:	Enter a zero in the EST column for each activity with no prerequisite
Rule 3:	Any activity with no prerequisite has an EFT equal to its duration
Rule 4:	Each activity with a single prerequisite has an EST equal to the EFT of the prerequisite, and an EFT which is duration, d, later For example: Activity B;duration 7; prerequisite A $EST = 4 \quad EFT = 4 + 7 = 11$
Rule 5:	Each activity with several prerequisites has an EST equal to the latest EFT of its prerequisites, and an EFT which is d later For example: Activity G; duration 5; prerequisites D, E, F $\quad\quad EFT_D = 19 \quad EFT_E = 26 \quad EFT_F = 21$ $\quad\quad EFT_E = 26$ is the latest of these EFTs Then $EFT_G = EST_G + d = 26 + 5 = 31$

Now determine the latest times, according to the following rules.

Five Rules for Latest Times

Rule 1:	Start at the bottom of the table, and complete the second two columns (LST, LFT) together
Rule 2:	Each of the last activities must have an LFT equal to the EFT for the overall project; therefore transfer this value into the fourth (LFT) column
Rule 3:	The corresponding LST is always the LFT minus d For example, for Activity E: LFT $= 26$ d $= 12$ LST $= 14$
Rule 4:	Transfer the LST so generated into the LFT column for each prerequisite activity For example, for Activity G: LST $= 26$; Prerequisites $=$ D, E, F Hence LFT $= 26$ in all three cases
Rule 5:	When more than one LFT is obtained for any given activity, retain only the smallest. Cross out the others

Complete the table of times for the project

Table 6.4 presents the output of this scheduling exercise.

Plot activities on to a Gantt chart

Having now calculated the earliest start times, latest start times, earliest finish times and latest finish times, you are in a position to plot it onto a Gantt chart in the same manner as presented in Fig. 6.3. In this case there is scope for seeing parallel tasks (Fig. 6.7).

Table 6.4 Activity times for UX sub-project for Wheelyfast

Code	Prerequisite	Duration	EST	EFT	LST	LFT
A	–	4	0	4	0	4
B	A	7	4	11	7	14
C	A	10	4	14	4	14
D	B	8	11	19	18	26
E	B, C	12	14	26	14	26
F	C	7	14	21	19	26
G	D, E, F	5	26	31	26	31
H	G	4	31	35	31	35
I	G	2	31	33	33	35
J	H, I	1	35	36	35	36

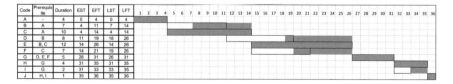

Fig. 6.7 Gantt chart for UX sub-project for Wheelyfast

6.1.12   Checklist for Time Management of Your Project

The following checklist is for you to use that helps you plan your time management for the project. You will need to adapt it to your own particular circumstances. The list of items is not exhaustive, and some of these may well be irrelevant. If an item is relevant, but you don't know the information, suggestions have been added regarding where to find it. You can also use the space to note where you should look. It is written in the first person as these are answers to questions you need to address yourself.

Overall hours

A: Overall hours = (check university's guidance documents/module descriptor)	B: Overall number of weeks =	Average hours per week $(A \div B) =$

Deadlines and significant dates
Insert the dates where you know the following: (cross out any that don't apply)

	Date	Where to find information	Order of occurrence
Final hand-in			
Competition event			
Competition entry			
Project proposal			
Risk assessment			
Ethical approval			
Interim progress report			
Log book			
Draft of final report			
Final report			
Presentation slides			
Oral presentation			
Reflection on project management report			
Artefact			
Poster			
Website			

List any additional, self-imposed deadlines

e.g. sister's wedding			
.			
.			

Turn back to the checklist at the end of Chap. 3 for dates of meetings with supervisor and project client.

Team meetings will be held at the following frequency ....
Starting on ....

Actions to be completed to draw up the time-plan (At * select and insert appropriate time interval.)

☐ I have agreed a project title and supervisor; if not, date by which to be agreed...........
☐ I have agreed aim and objectives
☐ I know what outputs are required of this project for assessment

(artefact/executive report/proposal/progress report/final report/presentation/poster/
...........) and the weightings for these are:
.... / .... / .... / .... / .... / .... / .... /

☐ Estimated number of hours/days/weeks to carry out background reading
...........*
☐ I know the methodology I'll be using
☐ Estimated number of hours/days/weeks to compose literature review ....*
☐ Estimated number of hours/days/weeks to compose method statement ....*
☐ Method statement has been agreed with my supervisor/project client/team
☐ Estimated number of hours/days/weeks to carry out method ....*
☐ Equipment availability has been established with ....

(supervisor/technician/technical manager/other)
Dates of availability of equipment between: ....(start) and ....(finish)*

☐ Lab/workshop space has been agreed with ....

(supervisor/technician/technical manager/other)
Dates of availability of equipment between: ....(start) and ....(finish)*

☐ I know the departmental procurement procedure; name of administrator..............
☐ Advised time to get approval for a purchase order ....*
☐ Purchase order(s) for capital item(s) have been processed in accordance with the schedule below:

Purchase order(s) for capital item(s)		Purchase order(s) for consumables	
Item description:		Item description:	
Supplier:		Supplier:	
Department's PO reference:		Department's PO reference:	
Raised (date):			
Approved (date):			
Sent (date):			
Delivery time for item:			
Anticipated date of receipt*:			
Actual date of receipt:			

☐ All appropriate dates (noted with an asterisk*) have been entered in my time-plan
☐ Estimated number of hours/days/weeks to produce artefact/other item ....*
☐ Estimated number of hours/days/weeks to compose progress report ....*
☐ Estimated number of hours/days/weeks to write up results ....*
☐ Estimated number of hours/days/weeks to write up analysis ....*
☐ Estimated number of hours/days/weeks to form and write up conclusion
...........*
☐ Estimated number of hours/days/weeks to format draft report ....*
☐ Estimated number of hours/days/weeks to proof read draft report ....*
☐ Estimated number of hours/days/weeks to review and amend ....*
☐ Estimated number of hours/days/weeks to make poster/website/other ....*
☐ Estimated number of hours/days/weeks to compose presentation ....*
☐ Estimated number of hours/days/weeks to rehearse presentation ....*
☐ Estimated number of hours/days/weeks to make other output(s) ....*

Additional considerations for group projects

☐ First team meeting has been scheduled
☐ Contact has been established with all team members
☐ Contact details are in my smartphone contacts
☐ Contribution of each team member has been established with distinct personal
objectives; if not, date by which to be agreed ....
☐ Key dates have been listed and discussed with the team
☐ Problems and risks to the time-plan have been identified ....
☐ Mitigation for risks to the time-plan have been agreed ....
☐ Shared, online project calendar has been created with all team members and key
dates added

6.2   The Project Plan and Resources

6.2.1   About Resources

The most important resource in your project is you, the student. You are going to
initiate and carry out almost all of the work, and certainly, without you, your project
will not be completed and reported.

Generally, dissertation projects are individual affairs, but there may be projects that
are done in a group; or your dissertation could be part of a bigger project undertaken
alongside a group of people with whom you need to cooperate. A student group
project can be a great learning experience, as everyone is undertaking the project as
part of their studies, and everyone is going to be assessed; so, people in the group
are mutually supportive and behave as a team. This is the ideal.

There will be other people involved in your project, as mentioned in previous chapters, but this is the point to consider their respective roles in your project. Your supervisor and client (if you have one) will also contribute to the project in the form of advice, specialist knowledge and intellectual property. If your project is part of your supervisor's wider research, you might be offered use of facilities in their lab. If you are working with an industry partner as your client, you might be required to carry out practical work only at their workplace, for reasons of confidentiality.

6.2.2 Defining Tasks and Understanding Equipment

Once you have worked out what you are going to do—what tasks are needed to fulfil the objectives of your project—then you can establish what you might need to borrow or buy.

Equipment for physical experiments can be very expensive, So, hopefully, what you need is available in the university, or exceptionally, with the aid of your supervisor (or client), arrangements can be made to provide you with access to what is most expensive. Really big, expensive equipment, like a large mass spectrometer or a dynamometer, will probably require you to book time on it. If this is the case, depending on the demand of the equipment, you are likely to be allocated a time-limited appointment on a specific date and time, when you are expected to carry out all experimental work. Naturally, you will need to make sure you keep the appointment and are punctual, and are fully prepared to complete whatever procedure is necessary. So, you need to make sure that you arrive at the equipment with adequate knowledge and a workable plan on how it will be used. It may help you to know beforehand what the make and model of the equipment is, in order to look it up online and read the users' manual about operation. In some cases, the university may require you to get some specialist training on the equipment. As a student facing a sophisticated piece of lab equipment (Fig. 6.8), no one will expect you to have seen it before, let alone operated it; however, the more you demonstrate that you have done your homework in familiarizing yourself with it before you encounter it, then the people-in-the-know are more likely to help you. If you have prepared well, then your enthusiasm will be evident in your understanding, or attempts to understand, the task ahead.

For some projects the equipment needed might be quite ordinary—for example, if you are carrying out a survey of engineering plant, make sure you have a camera, pen and notebook. Many people carry cameras all the time, in the form of their smartphone. Be aware when using it in unaccustomed places, like building sites, as it is more vulnerable than normal to being dropped, damaged and lost (or as one building services student did, dropped into a septic tank). Also, some locations may prohibit or have restrictions upon photography; if in doubt, check first.

Testing golf drivers

Fig. 6.8 Laboratory equipment can often be complex and expensive—and can't be mastered quickly and easily. (Illustration of a golf club testing machine by Mr William Heath Robinson)

6.2.3 Consumables

The other thing you might need, for experimental work or when you are creating an artefact, are consumables. Examples of these are paint, printer paper, trace dye, small electronic components and lubricants. Do not assume that these things will just be in stores for your convenience. It is far better to make a list and collect these things together BEFORE you start carrying out any work. Don't let the work evolve, then discover you don't have (or cannot easily get hold of) something you need to complete your work. Successful students anticipate what they are going to incorporate into, or consume during, their project.

6.2.4 Software

An important resource for many projects is software. Often, the dissertation project is an opportunity to learn a new software package that is widely used in industry, and so increasing your employability.

If your project needs specialist software, you should first check how it is available. Some software is freely available on the internet for a free subscription (i.e. the supplier just wants to collect your data), although this might not have all the functionality of the pay-for version. Alternatively, some software is available as a free or cheap "education edition" to anyone registering with a university email address, or proof-of-registration as a student (the suppliers are obviously hoping that graduates will persuade their employers that they need the—inevitably expensive—software in the workplace: for now, just take advantage of what's available).

Your university may have a license for specialist engineering packages, in which case you might need to find out where it can be accessed; for example, some specialist software may only be available in certain computer labs, as it has specific hardware needs, or the license may be limited to so many "seats". In these circumstances, you might want to talk to your supervisor about how dependent your project is on the use of this software, and whether you need exclusive access for a period of time—for example, if periods of processing are required for model simulation (the time and amount of computing power needed to run modern models can vary considerably: make sure you understand your project's specific needs, and whether the available hardware and software will be adequate).

6.2.5 Risk Management

For many students carrying out computer-based projects, your workspace is likely to be all electronic. If you are carrying out any experiment or testing you are likely to need laboratory space, or, if you're building something, you might need space in a workshop. If you're testing or building, talk to your supervisor about your space requirements and do not assume that, because they set the project, they have organized lab, or workshop, space for it: that might still be the student's job.

Be aware that if you are building, testing or experimenting, then you must manage the risks inherent in such activity. All universities will have their risk assessment process, and will expect you to undertake specific health and safety training within their facilities. It is good practice to include evidence of these things in your dissertation report. This form of professionalism raises grades, and is beneficial for discussing later in job interviews. It's likely that the senior technician in any laboratory can show you the notes for this and arrange sign-off of any risk assessment.

6.2.6   Technicians

If a project needs space in which to carry out the work, or use of any lab equipment, then it will probably also require the support of technical staff. Make sure that technicians will be available, and prepared to assist—whilst this is their job, they are likely to be supporting a lot of projects. If in doubt, always talk to the technicians themselves. Technician support in universities is a valuable, but often scarce, resource. Technicians are skilled and highly trained and educated professionals, who probably have been part of so much research over the years that they know both the processes and techniques that are in the textbooks, and workarounds that are not!

Good project students will make sure that they know which lab or workshop they are likely to be working in; have gone along and introduced themselves to the technician in charge of that lab; have discussed the aim of the project and ideas they have for carrying it out; got an inventory of equipment available to them; and booked a time-slot to learn how to use the most complicated piece of kit, before they have finalized their project proposal. Generally, technicians love students who show enthusiasm for their studies, and are clearly willing to learn. If treated with courtesy and respect, technical staff will bend-over-backwards to assist a student get the best possible results for their dissertation project (Fig. 6.9). Conversely, students who are not considerate of technicians may discover many difficulties in their path.

6.2.7   Administrators

During the course of your dissertation project, you are quite likely to come into contact with professional and administrative staff of your university, who you have not previously met before. This is particularly the case if you are buying things for your project. Procurement support is a very specialist, and highly regulated, area of business: both in universities and industry. If you are not provided with a guide to making purchases for your dissertation project, then your supervisor may be able to direct you to the right person to speak to, or you might need to talk to the technical

Fig. 6.9 You may often not be permitted to operate the more complex equipment without support from a qualified technician (by Mr William Heath Robinson)

manager of your lab space, or your course administrator. It is always advisable to find out who is in charge of procurement *before* committing to buy anything. This member of staff usually has an institution preferred-supplier list, and a great deal of experience negotiating the best deals. All you need to do is identify exactly what you want to buy, and from where it is available.

6.2.8   Vital Information

The last resource we'll consider here is information that generally available that relates specifically to your project. For example, data sheets for materials or chemical compounds, or user manuals for particular items of plant or equipment that form part of your investigation. Much of this type of information is now available online. Be aware, however, that sometimes the current details you're looking for are behind a paywall. If you encounter this, talk to your supervisor or subject librarian, as they may know a way of getting the information without incurring additional cost, or may be willing to make the payment for you from university funds.

6.3   Budget

6.3.1   About Budgets

For many people, their dissertation projects may not incur costs, or it might be as simple as a one-off subscription fee for a software package (cost = £50), and submit the report by electronic submission (cost = £0). If you have a project that involves making or building something, however, it is likely that you will be required to construct a budget and obtain financial agreement from both your academic supervisor and departmental budget-holder, before you can go and talk to the person responsible for procurement.

This process can be laborious, but it is there to ensure that all students who are entitled to financial support get it.

Alternatively, in some universities there may be no in-house funding to support student projects, and you to have to self-fund or seek out sponsorship. So, whether to monitor your own spending, of the university's or that of a benevolent sponsor, it is advisable to have a budget against which to work.

6.3.2   Relating the Budget to the Project Plan

The budget should relate to that list of tasks composed for the time-plan. Each task, or row in the project plan, may or may not have a cost associated with it, with respect to buying equipment, consumables or services. This is the point at which you create another column entitled "estimated budget". Fill this in as well as you can—probably drawing from freely available information on the internet. It's an idea here too to discuss with technical staff in your university, as they should have experience from previous projects of how much materials and processes cost (or ways to get things cheaper than the pricing you can see online). If costs are at this stage estimated, it's appropriate to round to the nearest currency unit (or 5 or 10)—there's no need to work with excess precision in financial estimation than there is in engineering calculations. The total of this column will give you and your funder an idea of how much the project is going to cost. This then needs reviewing, usually between the student and their supervisor, to confirm whether the budget is acceptable, or needs to somehow be reduced.

6.3.3 Spending the Money

Once the budget has been agreed as within the available rules and resources, it's time to start considering how spending should be done. Unless all the money will be your own, don't start buying things yourself thinking that you'll be simply reimbursed. Reimbursement is too easy to abuse, and it is difficult to demonstrate transparency. Hence, reimbursement is not a usual financial practice in the twenty-first century. Most organizations, including universities and most engineering companies, will also have ways of working designed to ensure that they are not unwittingly supporting illegal financial activities, such as money-laundering. Consequently, this will often mean that reimbursing students for purchases for a project requires administrative time and effort. So, to avoid time-delays in obtaining purchases, make sure you understand the procurement process. This may be as simple as knowing how to raise a purchase order, whether it is within your university or within your sponsoring client's company. To protect your own interests make sure you know the procurement system before you start buying stuff yourself—loading up your credit card or online purchase account, or emptying your bank account.

Often, you may need to obtain quotations (typically prices for the same items from at least three sources), or there may be somebody whose job is to do this for you. Often in the "real world" there's a threshold level of expenditure, above which competitive quotations must be sought: it's unlikely for a dissertation you'll be above this, but that doesn't stop it being good practice anyhow (and doing it shows an understanding of at least one aspect of engineering procurement). This process is called "tendering" and can sometimes add weeks to a time-plan, so it is important that you have an idea of how long this will all take as early as possible in your project.

6.3.4 From Estimated to Actual Budgets

Once you start getting cost information from quotations from suppliers, it is normal good practice to enter this information in your project plan in a separate column entitled "actual budget". As this is now actual rather than estimated, give the full and correct values—rounding is no longer appropriate.

While monitoring project expenditure, ensure you are not exceeding the agreed budget (and indeed, in most organizations underspend must also be accounted for and explained: that is of course the sort of environment the dissertation is preparing you for). This exercise in monitoring can also alert everyone involved when purchasing of important items has been missed; especially when ordering and delivery times are long, then that can be a very useful warning. If an industry sponsor is paying the bills, they are also likely to want to know accurately when money will be spent, as this is important for their own financial management systems—and again, useful practice for when graduates move into industry as well.

There may be occasions where this work shows that a project is going over budget. If this occurs, it's vital to discuss this immediately with whoever is responsible for authorizing payment (and if that isn't them, also with the supervisor). Sometimes budgets can be increased (but this is likely to require formal justification) but sometimes they cannot, in which case action will be needed to keep things from overrunning, financially (Fig. 6.10).

Fig. 6.10 The university accounts department: hard at work supporting student projects (by Mr William Heath Robinson)

Chapter 7
Designing the Dissertation

Abstract This section describes how to design major types or components of dissertation projects. These include the feasibility study, experimental studies (both characterization and hypothesis testing), product design activities—including the use of Figure of Merit calculations, and the design and use of surveys to gather data in support of engineering projects. For the latter also, the importance and general mechanism for obtaining ethical approvals is described, as well as various ways of obtaining, storing and reporting data.

© Springer Nature Switzerland AG 2020   95
P. Gratton and G. Gratton, *Achieving Success with the Engineering Dissertation*,
https://doi.org/10.1007/978-3-030-33192-4_7

7.1 Designing Feasibility Studies

7.1.1 What is a Feasibility Study?

Feasibility studies, which are used in all branches of engineering, are a process of assessing how practical it would be to carry out a proposed plan or method. For example, the asset manager for a housing association might want to know about replacing old housing stock for new, and how many new-build homes could be created on a site where there are currently few residents, without increasing annual energy costs. Or, the Chief Engineer of an airline might want to know the cost and complexity of upgrading all the cockpits of a particular aircraft type to the latest technology displays and navigation equipment.

Feasibility studies are also important in guiding financial decisions and planning—and much can be learned about them from business management textbooks. However, in engineering, we will certainly do not have cost as our *sole* consideration. Authors of feasibility studies must consider what the priorities are—for example, basic achievability, regulatory compliance, safety to human life and health, efficient use of energy and resources, minimizing waste or harmful by-products, reducing mass or increasing acceleration, and so on. Feasibility studies must always, of course, satisfy their name—demonstrating whether something is actually feasible and, if so, how it would be done: including the technical and regulatory risks attached to particular means of delivery.

In the context of a dissertation, the feasibility study may be the first part of the dissertation, or (perhaps more likely) it may be the entire dissertation project. Industry feasibility studies can sometimes use hundreds of engineers and millions of dollars, so there is nothing "just" about a dissertation only consisting of a feasibility study. Equally, if it is "only" part of a bigger project, it is very much an important part, and should not be skipped over.

7.1.2 Defining the Feasibility Study Topic and Questions

Consider this possible opening title for a feasibility study. "Evaluate the feasibility of converting the Students Union minibus to all-electric, using student labour and readily available parts."

The first thing that is obvious in reading this is that the feasibility problem needs defining much more tightly than the short sentence above. For example:

(1) What's the budget, man-hours and financial for the feasibility study? The study itself will then include determination of costs for the full project, which is likely to include cost variations and determination of the time and costs of a full delivery.

(2) How do we define "readily available" parts? Presumably, some definition along the lines "can be ordered for delivery to campus within 1 month".

(3) Student labour may vary from former vehicle maintenance apprentices, and final year automotive design and electrical engineering students at the highest skill level, to first-year humanities students at the lowest. You need to know how many students are prepared to participate, at what skills levels, and hence, how many labour-hours will be contributed?

(4) Is there a deadline for conversion, and what period may the minibus be "offline" for work to be done to it?

(5) What performance is required from the minibus, in terms of range, minimum acceleration, maximum speed and so on?

(6) What extra infrastructure (presumably a charging point in an appropriate place, at the very least) also needs to be in place?

(7) What statutory regulations must be complied with in order to be able to run the bus on the public highway?

The first task of the student (or any engineer leading a feasibility study) is to ensure that the scope and resource of the feasibility study is well defined. Sometimes, this will be laid out by an expert customer from the start, more often detailed discussions (between the student/engineer, supervisor and client) will be needed to agree these details.

7.1.3   Designing and Commencing the Study

As with almost all engineering projects, feasibility studies must commence with a literature review and determination of a workable project plan via critical path analysis and a Gantt chart. The only major difference between this and many other engineering projects is that almost all of the components will consist entirely of labour, although there may still be requirements for small elements of computer-intensive analysis and experimental work (planning our electric minibus, for example, testing the characteristics of a single example of a preferred battery type.)

The approach to be taken is not unlike any other research project—the output must be a set of conclusions and recommendations, built upon a clear narrative flow from start to end. There will be uncertainties in those conclusions—rather more so than in most research, because it is in the nature of a feasibility study that it is making predictions about something that has not yet happened (and may sometimes never happen—it is very common for a feasibility study to be undertaken in order that a go/no-go decision can be made about a proposed project, and in doing so, this is very much real-world engineering).

In designing the study, a good practice is to start by fleshing out the structure of the eventual report, and you may find it helpful to start with the titles of the primary conclusions and recommendations. Working backwards from those towards the opening literature review should show the information that will be required to form the conclusions. That will show the mechanism by which the conclusions can be drawn: which will be some combination of literature review, survey, analysis, design and experimental or trial manufacturing work. Once you've worked out the ideal combination of these things, you can line this up against the genuine total resources you'll have available for the project. Almost certainly meeting the available resources will then mean scaling back all of the ambitions for the amount of work that can then be done—but it should then become possible to create a workable plan.

7.1.4 The Ongoing Writing Journey

The primary output of a feasibility study is the report; so, it is important throughout the project to keep revisiting that report. So a plan that involves a linear journey of plan→literature review→experimental work→analyse→write report won't work. Writing has to be a fundamental part of the process of conducting a feasibility study, and an initial draft report; however basic, ideally will exist from very early on in the process, and be continuously developed through the project. If there is a client, it's likely that they should be involved in that process, at least to the level of agreeing that the right questions are continuously (and consistently) being pursued.

Plans must allow for the considerable time involved in this writing process—but conversely this should mean that the final writing phase will require less time than for a design or research project.

7.1.5 What is Not in a Feasibility Study?

What's important to remember is that a feasibility study exists to identify what can be done, how, and the resource (money, time, people, skills) required to carry out that task—along, ideally, with some consideration of the risks involved. When we describe risk, this includes safety issues, of course, but particularly means "technical risk"—the potential for variations in the resource requirements, and any significant potential for the project not to be achievable.

Any design, analytical and experimental work should always be constrained to only that required to determine the conclusions and recommendations of the report. This does not mean that none should be carried out—but, for example, if a bridge would require a thousand similar components to be designed and analysed, it would be reasonable for the engineer conducting a feasibility study to perhaps design two components, so that the average time and computing power required can be estimated from practical experience. Outline design work, or computer analysis is also entirely in order—for example, to estimate the amount of steel needed for a bridge, or the likely performance of an aeroplane—but this must always be at an outline level. In reality, the resources available to a feasibility study project will be fixed (whether in a dissertation, or subsequently in your engineering career), so the quantity and quality of supporting analysis will almost certainly be capped by the budget and delivery deadline. Any feasibility study will, ultimately, be doing the best job it is capable of doing within the resources allocated to it.

7.2 Designing Experiments

7.2.1 About Experiments

As a science and then engineering student, you'll have carried out, and written up experiments. They are a fundamental part of all science-oriented subjects—you can't determine everything by analysis or pure observation, and sooner or later someone needs to actually create experimental results to test theories, inform models, provide values or establish physical processes. However, curiously enough, whilst most students learn how to perform and report experiments, generally how to design experiments is seldom taught very well. So, at your dissertation stage, if it does have an experimental component, you may be have to learn for the first time how to construct your own—sometimes complex—experiments. The good news is that, whilst often complex, this can also be very rewarding.

Not all dissertation projects will require experimental work, but many in engineering and the science stream will, so it's important to understand how to design and conduct your own experiments.

7.2.2 Defining the Question

The first and most important part of designing an experiment is to work out what question you are trying to answer. A typical question may be like one of the following:

– What is the ultimate tensile strength (UTS) of this material?
– What is the stalling angle of attack α_{max} and lifting coefficient $C_{L.max}$ of this aerofoil?
– What is the calorific output of this fuel when burned?
– What is the torque:rpm relationship for this engine?
– What is the human response to this alerting information?
– How much air leaks from this building?

You'll notice two common characteristics to all of these questions. One is that they all require the experimenter (some industries will call them the *test conductor*) to take measurements, and the second is that very rarely will the experiment result in itself be enough to solve an engineering problem—usually it will be a necessary contributor to eventual engineering conclusions. For example, you measure a material's UTS so that you can then use that data in predictions of structural failure, and thus to ensure an adequate design. You want to know α_{max} and $C_{L.max}$ for an aerofoil so that you can then design the configuration of a machine such as an aeroplane or windmill which uses that aerofoil section; you want to know the calorific content of a fuel so that you can predict the performance of a machine or process that burns the fuel and so on. So we need to know what the "exam question" is, and —very importantly—how precise and accurate that result needs to be. It is this requirement, for an answer with particular qualities, to feed into some other purpose that must be used to define any experiment. It's never good enough to simply say "we'll test this to destruction", or "we'll put this in the wind tunnel".

7.2.3 Do You Need a Hypothesis?

A hypothesis is a form of informed guess, and some science experiments do need one. For example, if we are trying to determine whether changing from single to double glazing reduces energy losses. Many texts will say that a hypothesis is essential to design an experiment, but this is clearly untrue, as many experiments are there to quantify something, not to test whether it exists or not (you can get picky and say that the hypothesis is that there's nothing there—but clearly any object has a mass, we don't need a hypothesis suggesting that it doesn't; we can just measure it).

So if the question is: *"Is there an effect?"* of some form, then yes, you require a hypothesis. If the question is *"what are the characteristics of?"*, then no, you do not require a hypothesis. Clearly, it's important to know whether you have a hypothesis, and if so, what it is.

7.2.4 Control, Dependent and Independent Variables

Any experiment will be defined by three types of variables, which are:-

Control variables: these are the variables (also in some fields called *common data*) which are left unchanged throughout the experiment. It's always important to keep a significant proportion of values or factors unchanged throughout an experiment so that you can determine whatever it is you're trying to find out. With the questions above this may be the test machine, the temperature and pressure in the wind tunnel, or the design of a calorimeter setup.

Independent variables: these are the things that are changed in order to run the experiment. So, with the structural strength experiment it may include the load, the shape of the test item or the material it is made from. With the wind tunnel experiment, it may include the air velocity or the angle of attack.

Dependent variables: these are the quantities which change during the experiment, and which are measured or observed. So, this may be the strain or failure load of a component, the velocity or angle of attack at a wind tunnel test item, or the fuel that's used in the calorimeter.

Deciding which variables are control, independent and dependent variables is important in the early stage of your experimental design.

7.2.5 Now Design Your Experiment

So, you know what your question is: whether you have an associated hypothesis—and if there is one, what it is. You should hopefully have a good understanding—from your earlier literature review and ongoing dissertation work, and all your previous lab experience—of what you want to measure and how. Simply going to the lab and getting on with it (even if the technicians will let you!) is seldom going to work well. You need a plan! Best practice is to formally record this—and probably get your supervisor to go through it with you and discuss or help you correct the critical points. It is usual to obtain your supervisor's approval to proceed with experimental work in any case—and they can only approve a clear and unambiguous plan. Usually you should also need to submit a risk assessment of your experimental work before you can proceed with it, and sometimes gain ethical approval before you can proceed.

The complexity of your plan will depend very much on the complexity of what you want to achieve, but a good structure is normally like this:-

Introduction
This will normally cover administrative information such as author, their supervisor, the test item location and (approximate) test dates, as well as a general introduction to the reasons the experiment is to be carried out including, if there is one, a statement of the hypothesis.

Specification
A more detailed statement of the test item(s) and test equipment to be used.

Test plan
Usually this section consists of a few explanatory paragraphs of the setup and general conduct of the experiment, then (usually in tabular form) details of the actual experiments to be carried out and (very critically) the data that will be recorded during the experiment. Good column headings are along the lines of serial number/test to be carried out/test conditions/data to be recorded. It does no harm to record excess data (a particularly useful addition is, for example, to take lots of digital photographs during your lab work).

Whilst writing the test plan section, mentally rehearse how you would actually conduct the experiment. Try to design your experimental procedure in a manner that allows you to get everything done efficiently, and reflect that in the layout of the test plan. At the same time, write in flexibility: as you progress you may find that lots of results in a particular range of conditions are the same, so you can go through them quickly, or that in another range something unusual or interesting is happening and you need to take longer. In addition, you'll (hopefully) find that as the test conductor, you become more efficient. So do not unnecessarily constrain your plan to a very narrow method. Also know how much time you can afford (or will be permitted) to conduct the experiment and make sure that your plans are compatible with that.

Risk assessment
This is usually required, but check with your university's rules. Most countries and working environments nowadays require formal written risk assessments for activities such as laboratory experiments; however, this may be covered by existing risk assessments in place, and even if not may need to be in a separate document anyhow. If your university does require a risk assessment, it will definitely have a required layout, or *pro forma*, so don't make this up.

Ethical assessment
Again, whether this is needed or not will depend on circumstances. It's increasingly necessary to have a formally signed-off ethical approval for any work involving human beings (e.g. human factors experiments, interviews, work with assistive devices for disabled people), or with animals (e.g. designing farm feeding devices, and certainly any animal experiments—although they are obviously rare in engineering). Similarly, ethics and data protection now mandate very careful management and consent where any personal data is collected—particularly in any EU or former EU country, under the General Data Protection Regulations (Fig. 7.1). There should be clear local rules about this—make sure you understand them, who need to sign-off ethical approvals and whether this needs to be a separate document or included in your main experiment plan? Be aware that it's not unusual for universities to refuse to mark assignments, or even to award degrees, if an ethical assessment was required, but not done before the experiment was carried out; undoubtedly, this will soon be the case also with GDPR breaches.

7.2.6 Writing Up Your Experiment

Check with your supervisor whether there is an approved standard layout for experimental work. If there is, use it. If there isn't, be guided by the format of lab sheets previously used in assigned experimental work in the past. If you are in doubt, be formal and to the point in your layout, and don't be afraid to state the obvious. Typical headings may be:

- Title
- Author
- Location
- Overview of experiment to be carried out
- Equipment required
- Dates for planned tests
- Order of experiment

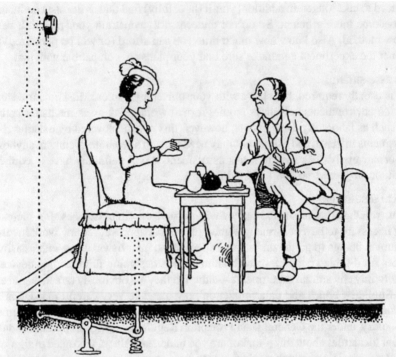

Inoffensive method of ascertaining the weight of a lady friend

Fig. 7.1 Now unacceptable metrology practices from a bygone age (by Mr William Heath Robinson)

- Minimum data required
- Reporting requirements
- Risk assessment

You will find it impossible to write these sections well sequentially. The best way is to open a word-processor document with the right headings (tweaked as required for your local needs) and effectively write them in parallel. Whilst it may cause some personal confusion, by jumping back and forth between the headings, you can ensure that they're self-consistent and make sense from the perspective of your experimental objectives.

Now, if you have time, the best thing to do is to start writing your lab report. Just because this is for your dissertation, it does not give you a reason to skip this vital part of the process—you want that thorough report available to you as you create the dissertation report, written in pretty much the same way as all of your student lab assignments from earlier on in your degree(s). While starting to write it, before you go to the lab and start obtaining data you will ensure that the experiment plan

A WELL·THOUGHT·OUT
AND NEARLY SUCCESSFUL
EXPERIMENT BY EARLY
RAILWAY PIONEER ·
VIGNETTE .

Fig. 7.2 An early railway technology experiment, illustrated by Mr William Heath Robinson

is good enough to let you write the report you need. Very often you will find that by working in this way, you find deficiencies in the plan in good time to go back and fix them. Your lab report should usually follow the format you've used earlier in your studies: (introduction, method, conditions, results, conclusions, recommendations)—or something very similar. In particular, do think about how you'll do your data analysis: not just in your immediate lab report, but any subsequent analysis and incorporation of lab reports into your broader analysis that goes into the dissertation. Are there enough data points for statistical significance (if that's an issue), and do you have all the data you'll need for every calculation you'll be doing?

Now you can go and do your experiment (Fig. 7.2), and write your report up. This report will be an essential resource as you prepare your full dissertation report.

7.3 Designing Products

7.3.1 What is Product Design?

Product design is classically about three components: form, fit and function. So to take a simple device—a drink bottle, the form is the visual appearance, shape, colour, texture, appearance and content of the labels; the fit is whether it carries the right amount of liquid, fits into machinery used to fill it, and into bottle storage and transfer spaces; and the function is the ability to be filled and poured, to retain liquid under design conditions, and possibly more complex requirements such as manufacturing methods, and reuse or recyclability.

You may find that a great deal about product design is about the "form" aspects, since *graphic design* as a subject has a strong concentration on the aesthetic. *Industrial design* tends to give more consideration to fit and function. However, as engineers, we must give significant weight to all the three components—indeed, in later professional practice you might well be guardian of "fit and function" so that a product designer, architect or similar professional can concentrate primarily on the aesthetic aspects. For the purposes of your dissertation you will, however, almost certainly need to balance all of these fundamental needs, and excluding any major aspect will attract penalties.

7.3.2   Defining the Product Design Specification

It is highly likely that you'll be required to use a specification as part of any design study. You may be given, or even in some fields, an initial conceptual design (this is common, for example, in aeronautical engineering or building design); alternatively, you may well be undertaking the background research to determine your own specification. In doing so, the fundamental need is that you clearly and rigorously define what you're doing and why, by reference to clear criteria. It would be highly advisable when you believe that you know what your design requirements are, to establish that in brief, unambiguous written format and discuss those early with your supervisor and client. Be guided by them about the format of that—there are too many variations in different engineering sub-fields for us to give you a template here: but useful pointers include:

- **Avoid vague terms** such as "large", "fast", "low speed": instead "25–50 tonnes", "over 150 m/s", "1–5 fps" are much more engineer-friendly.
- **Ensure the terminology you use is clear** and well defined.
- **Define standards**, when using them, including version numbers. Indeed, in general, don't be afraid to refer to authoritative documents (which for design specifications usually mean engineering standards rather than research papers).
- **Do not tie down requirements more than necessary**. In the above, for example "25–50 tonnes" gives much more flexibility than "37,500 kg" when the detailed design of something is not yet established, unless there's a very good reason to be so precise.
- **Use tabular format** for information where you can—it's usually much easier to refer than to lengthy text.
- **Ensure** your project client, if you have one, agrees with the specification. This may require you to work with them through several iterations until you're sure that all the three—what they want, what you can deliver and what your supervisor agrees—meet academic requirements.

7.3.3 Following the Design Process

You are likely to have been taught specific design processes on your degree course(s) and these should be followed usually by reference to the specific documents in which those were laid down. If you've not been taught specific (or multiple) design processes that apply well to your project, then consult particularly with your supervisor about their preferences, else research and declare your preferred approach and agree it with them. There are some important principles within that which you can usefully follow:

– As engineers we work with facts, generally expressed numerically. If you *must* work with opinions: take great care to ensure that these are well quantified. A good basis for this, for example, is that test pilots and aeronautical engineers use the Cooper-Harper scale shown in Fig. 7.3 to quantify opinion; using a diagram like that has much more weight than a simple "assessment on a scale of 1–9" that is far too often used and extremely subjective.
– Component selection is not the same as shopping. It's easy to go through catalogues and declare "this component met these criteria best, so was selected": there are few marks for this, particularly in an engineering dissertation you are expected to display use of analytical techniques in your work. See the section below on engineering figures of merit.
– You would have prepared a project plan, and be expected to either adhere to it, or amend it as you find it no longer works. This particularly applies to design activities, where the various parts of that process need to be planned also and fit into the overall project timeline as they're executed.

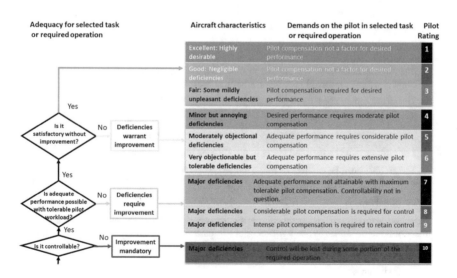

Fig. 7.3 The Cooper-Harper Rating Scale (derived from G. Cooper and R. Harper. The use of pilot rating in the evaluation of aircraft handling qualities. Technical Report TN D-5153, NASA, April 1969)

- If you are using material, process or component characteristic data, try to use actual verifiable data, not indicative data from textbooks or catalogues. Also, unless this is a specific requirement, don't design using *unobtanium*: a catch-all term for obscure materials made in tiny quantities in a laboratory and reported in research papers, but not actually available for manufacturing purposes.
- If you'll be prototyping, design for manufacturing techniques that are actually available to you (and talk to the technicians to make sure that this is really true, if the machine will be waiting for the next 6 months for a spare part—you need to know this now!).

7.3.4 Using engineering Figures of Merit (FoM)

The principle of figures of merit is putting a numeric score onto selection of a material, component or engineering solution. It does not remove *engineering judgment* from the process, but it does shift the location of that judgment to the construction of selection algorithms, and not to forming difficult to justify opinions about either inputs to algorithms, or judgment of their outcomes. The method is useful, of course, in general engineering practice—but especially in a dissertation project where demonstrating your ability to use analytical methods is particularly important.

Here's a (simplified) real-world example that one of the authors of this book did whilst helping a charity select the best aeroplane for delivering medical support in West Africa:

Introduction

The following were taken as absolute requirements:

- The aircraft must be of a type, and in particular use an engine type, for which product support will be reasonably readily available in West Africa.
- The aircraft must use a fuel type readily available in West Africa.
- Ability to use short runways.
- The aircraft must be capable of single- or-two pilot operation.
- The aircraft must be certified either in the USA or Europe.
- The aircraft must be capable of flying above 20,000 ft.

Value judgments

Selection of an aircraft for a practical role requires consideration of the purposes and assignment of numeric values to allow meaningful comparison of aircraft types. For this exercise, the following values are considered most significant:

(1) Maximum range
(2) Useable payload at maximum range (all aircrafts will have increased payloads at reduced ranges)
(3) Take-off distance required (TODR)

In order to try and define a "best" aircraft of the set considered, this was done by calculating for the set a figure of merit (FoM) as follows:

$$\text{FoM} = \left(2\left(\left(\frac{\text{Payload}}{\text{Max Payload of set}}\right) \times \left(\frac{\text{Range}}{\text{Max range of set}}\right)\right) + \frac{\text{Min TODR of set}}{\text{TODR}}\right) \times 33\frac{1}{3}$$

This will give a score out of 100, which compares to the best aircraft characteristics within the set, with a score of 100 being theoretically achieved by an aircraft only if it simultaneously met all of those best characteristics. The formula is set to effectively give primary weighting to (payload x range), with reduced emphasis given to the importance of take-off distance.

Aircraft types and numeric analysis

A number of aircraft types were considered (see Table 7.1) . All figures are taken from publicly available sources and should be regarded as indicative rather than absolutely accurate.

Table 7.1 Details and comparison of aircraft types

Type	MTOM[a] (kg)	Fuel capacity (kg)	Typical empty mass (kg)	TODR[b] (m)	Approx max payload with full fuel (kg)	Nominal range (nm)	FoM
Single engine							
Quest Kodiak 100	3,290	969	1,710	458	836	1,113	*62*
Pilatus PC12	4,740	1,226	2,800	808	454	*1,560*	42
Cessna 208 Caravan 675	3,629	1,009	2,155	626	481	946	37
Cessna 208 Grand Caravan	3,985		2,291	738	685	917	40
Cessna 208 Caravan Super Cargomaster	3,969		2,183	762	793	869	42
Pacific PAC 750XL	3,395	775	1,410	*364*	*1,124*	582	58
Piper PA46-500TP	2,309	527	1,544	743	268	1,000	27
Twin engine							
Viking DHC-6-400 Twin Otter	5,670	1,172	3,400	366	1,012	920	*69*
GAF Nomad N24A	4,263	834	2,150	360	521	840	50
B-N Defender	3,855	908	2,130	565	724	1,006	49
Dornier 228	6,660	1,845	3,739	524	990	808	54

Best values are shown in italics
[a]Maximum take-off mass
[b]Take-off distance required, to clear a 15 m obstacle, at standard sea-level conditions and MTOM

Conclusions

Taking a FoM cut-off of 55, the three most suitable aircraft currently available for the West African role, therefore, appeared to be the Viking DHC-6-400 Twin Otter (typical price circa US$1.5–3.0 m for a 30-year-old aircraft), the Quest Kodiak 100 (typical price circa US$1.4–1.7 m for a 1-year-old aircraft) and the similarly priced Pacific PAC-750XL. It was hard, however, to justify the substantially greater purchase cost of the DHC-6.

Considering the two aircraft types, the decision then became one of range versus payload. The PAC-750XL had the superior payload and better short-field capability, but a limited fuel capacity which restricted effective range to about 580 nautical miles. The Kodiak had a lower payload by 25% / 288 kg, but greater fuel capacity which resulted in 90% better range.

The example above does not give a definitive answer, nor does it absolve either the engineer using the method, nor their client from having to make decisions themselves. The engineer decided, in discussion with their client, on basic criteria (number of engines, fuel type, size range, number of pilots and service ceiling) and also decided on their selection algorithm (a balance of payload, range and take-off distance). However, once having made those decisions, the actual analysis was neutral, using verifiable values—it was also done in a way that would allow questioning of the method, and thus recalculation with different balancing of input parameters. All this is ultimately healthy. In terms of the output, sometimes a FoM method gives a very clear and unambiguous result, but (as is often the case) here it did not. It presented multiple "good" outcomes for consideration and further evaluation by the client. At that point they are likely to apply additional criteria—for example, compatibility with the support infrastructure for other aircraft already in use, availability, running costs, ease of recruitment of qualified pilots and so on. So, FoM as a decision method is normally only an interim stage in making design decisions, and does not absolve engineering judgment and calculations.

This method can be readily extended to almost any selection problem—for example

- what material to make a support bracket from,
- what type of glazing to use in a new building,
- electric motor and battery selection for anything from a small robot to a bus.

Similarly, there will invariably be further "judgment" decisions to be made informed by the FoM method—in these three cases, for example:

- based upon what manufacturing processes are available,
- the skillset of the workforce, and
- existing supplier relationships.

Other factors may often be introduced (cost, material properties, delivery times, other performance factors, ethical considerations etc.) at any stage, making this method necessarily flexible.

Therefore, use of FoM is a valuable method to include in your design processes;it enables good outcomes and allows a student to show competent use of analytical methods. But it is still only as good as the skills of the engineer using it.

7.3.5 *Presenting Your Design*

How you present a design will be both discipline and project-dependent. You should normally ensure that you do the following:

– Use formal presentation (e.g. technical drawing) methods that you've been taught to the highest standard you can.
– Don't be excessive with the amount of fine detail you include in a dissertation report. If you've designed a reasonably complex machine, then there may be tens or even hundreds of drawings—a representative sample—with good explanation is quite adequate. Of course, apply the same high standards to all of your work, but don't necessarily ask for every part of that to be assessed.
– Do show that you've followed a good formal design process, as taught or encouraged by your institution.
– Also use explanatory schematics or sketches so long as they are clear and informative.

When you present a design make sure that you do so in the overall context of your project. Explain whether the design is the entire project, or simply a means to an end (e.g. it was the project to design something to measure water flow, or actually to determine the flow through a canal). Ensure that you apply a proper sense of perspective and don't bury your reader in detail that they don't need. In most cases it's likely that the primary presentations will be (a) about the quality and nature of the design, and (b) the extent to which it meets the design specification. Consider these things!

Also, where you present the design—explain why you are doing so. Describe how it contributed to the overall project, or how your design meets a specification, or whether it is an interim stage to gain approval to manufacture. Each of these needs different levels and types of detail to be presented—if in doubt, talk to the supervisor and, if necessary, to the client.

7.3.6 *Prototyping*

Building a prototype of your design can be very satisfying, and also often very useful. However, it should be remembered that compared to what goes on in the greater world of engineering your time and available resources are very limited: your supervisor knows this and is not expecting you to magically create a finished and marketable product in the vast majority of cases. What you are likely to be doing is presenting a capability, demonstrating that something will work, and of course demonstrating your own abilities. So, be realistic in ambitions when prototyping—what is the real and realistic objective? You may be making a representative shape from building bricks or 3-D printed plastic parts, you may be demonstrating the basic functionality of a piece of software or you may be making an electronic breadboard in a project box. What's important in any case is to establish and get it right:

(1)   What are you really trying to demonstrate with your prototype?
(2)   If this is being assessed, make sure you know by whom and against what criteria.
(3)   What doesn't matter? (e.g. external aesthetic, long-term durability, suitability for mass manufacture) and if these things don't matter, you're not wasting valuable time addressing those aspects (Fig. 7.4).

Fig. 7.4 In prototyping, the objective can be just to demonstrate feasibility and basic principles—elegance and the correct scale can come later (by Mr William Heath Robinson)

7.4 Designing Surveys

7.4.1 Survey as a Means of Gathering Data

We're going to concentrate on surveys of people in this chapter. If you are studying construction engineering or something allied, then surveying is something completely different, and involves theodolites and levels for taking measurements. The type of surveys we're going to address here also involves establishing what the present condition is of something, with a view to carrying out a project that will cause change, but the focus is on people, not buildings or equipment.

The word *survey* is an umbrella term for a variety of methods for gathering data from people. Under this term, we include questionnaires, interviews and focus groups. We will also consider product testing, where the focus is on the user experience (UX). There is a tendency for engineers to underestimate the amount of work involved in surveying people for their opinion. There are many specialist books about this topic, especially dealing with the psychology of asking and answering questions. We recommend that you consult one or more these to gain the best possible understanding, and use this to justify your survey and its questions. If nothing else, you should look up for various resources that quote the "Hawthorn Effect", which was coined by the Australian-born industrial researcher, Elton Mayo (1880–1949), when he was investigating the effect of varying illumination on productivity in a transformer factory in Chicago. The lesson is about unintended consequences.

Survey design is an iterative process—like all forms of design. The design cycle is usually like that in Fig. 7.5. If you think that your objective is going to be best met by a survey, then you have to decide who you are going to survey and how you are

Fig. 7.5 The cycle of survey design

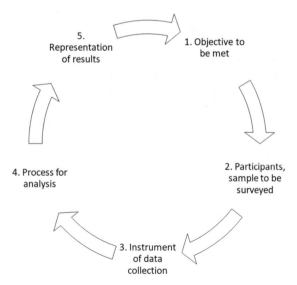

going to collect information from them, and process that information (analysis) and convert the analysis into a meaningful representation to meet the objective you have set.

The first time you go round the cycle, you will get an outline design of your survey. With each iteration of the cycle, you will find yourself refining the design a little bit. The challenge is going to be knowing when it is good enough to administer. Soon after you have the outline design, and long before you use the survey, you must consider the ethical implications of your survey.

7.4.2 The Need for Ethical Approval

Before we dive any deeper into the subject of surveys, the need for ethical approval must be addressed. This is usual whenever work is proposed that involves human beings, or animals of any kind (or parts thereof). Your university or department will have someone who is designated ethics officer, and probably there will be an ethics committee with representatives from all academic departments, including engineering, who will have the job of considering all the applications for ethical approval throughout the university. Note that there might be a delay of days or weeks to gain ethical approval. It is important for your schedule that you find out what is the lead-time—like your university's procurement process. Most institutions make it as simple as possible to gain advice about the ethical approval process. They are not there to stop you carrying out your research but to ensure that no harm befalls anyone in the process of carrying out your project. If your application is rejected, it is usually on the basis of insufficient information—and the representatives are usually very keen to help you get together the information needed.

There are several reasons to need ethical approval—personal safety, safety of your subjects, reputational safety. The personal safety aspect is one that is easily overlooked in the enthusiasm for a new project. Several years ago, a young man wanted to measure a cross-section of the population as they wore a harness to gather data in support of assessing its efficacy as a restraint. He was keen to measure a balanced sample of men and women, as far as possible. It was pointed out to him that it would be prudent to have a female colleague with him when measuring female subjects, not to suggest that he might do anything to them, but to protect himself from potential allegations that he had. Another student wanted to look into the employment arrangements in the construction industry. It wasn't until after he submitted the report that anyone realized that he had collected the data by hanging around building sites at the end of the working day, asking the workers to take part in his survey about pay and conditions. Fortunately, no one took exception to his asking these rather personal and commercially sensitive questions, but it could all have gone much less well.

The safety of participants in a survey is an obvious concern. No one would volunteer to participate in a research if they thought they were going to get hurt… or would they? There are plenty of old papers in psychology that detail tests involving electric shocks, and privations of something vital to sustain health and well-being. We once had two students who proposed to test factors that affected a pilot's visual environment, one of which was glare. Part of the experiment would require a bright light to be shone into the participant's face to simulate either flying into the sun, or getting reflections off objects in the cockpit. Being considerate of the time required for normal sight to return after being subjected to the simulated glare (the medical textbooks were very vague on the matter of how many seconds might be needed), taking ethical advice they designed the experiment to have the physical test first, and follow-up with contextual questions so that their participants would not be leaving their test room to climb straight into their cars or aeroplanes.

Guarding against reputational damage may be a consideration. Your dissertation project might be the first time you have ever considered surveying anyone, but imagine if every student were to decide to carry out a survey. One of the things your university will want to ensure is that the same group of people are not being surveyed repeatedly. Obviously, another one is that the researcher will be doing something new—and not merely repeating something that another student did last term. Sometimes, there might be good reason to repeat a survey in order to evaluate the effect of changes over time upon something or other. If this is the case, you need to explain in your application for ethical approval that you are consciously tackling this issue, and fully justify why it needs re-consideration and your piece of research. Finally, your university wants to protect its reputation by not being seen to condone anything illegal, (perhaps) immoral or just annoying. This latter point is a quagmire of liberal challenge, as universities are generally bastions of free speech, but that does not make them amenable to "anything goes in the name of research". So, certainly within the European Union, for example, you are not going to be allowed to research into humane means of judicial execution.

Your supervisor will be able to guide you as to whether your project needs ethical approval. Generally, it's needed when living things are involved. Talk to your supervisor.

7.4.3 Consideration of Legal Compliance Regarding Personal Data

In addition to ethical considerations, the ethical approval process of your university is likely to trigger questions about the legality of your survey proposal. One of the key aspects of this is the nature of information to be gathered. There will be a compliance manager in your university to ensure that what you propose will not infringe data protection laws, such as the European General Data Protection Regulation (GDPR), which might put your university's and your own reputation in danger. To help with this you need to be clear about how you are going to treat issues of confidentiality and privacy. Treating participants anonymously means more than not using their names. You also need to guard against using personal characteristic in descriptions of them that makes them easily identifiable.

Your institution's compliance officer will also be able to advise what needs to be considered when proposing the use of human tissue in experiments. These are generally specialist biomedical projects, and require a specific kind of registration with your university—probably not a factor in most engineering disciplines, but it will affect some people.

7.4.4 Design of Survey Questions

It is important that you justify each question in your survey—and don't burden your respondents with questions that are irrelevant. It cannot be stressed enough that your questions have to map directly onto your objective(s).

There are many types of questions, and it is usually best to include a selection of different types. Fundamentally, the two main types are *closed* and *open* questions. Closed questions can usually be answered with a single word or phrase. You can anticipate what the answer will be and offer a list for your respondent to select from. Questions that require "yes" or "no" are the simplest of this type. The next most common closed questions are the ones where you can use multiple choices—where only one of the choices can be valid to the exclusion of the others. Then there are lists where you ask your respondent to select as many items from the list as are relevant to their circumstances.

Open questions require your survey participant to frame their response in their own way, using their own words. The knack with these types of question is to provide a way to capture an adequate response, without letting the respondent give you a 13-page essay or one-hour interview with lots of repetition.

Whether posing open or closed questions, take care not to lead your respondent to a particular answer—this will compromise the validity of your results and conclusions. If you are making assumptions, be clear about the assumption, and permit your participants to disagree. For example, in a focus group to establish how aware the public are of safety measures in fairground rides, it was proposed to ask the following question:

> Explain a time when you were afraid for your life or severe injury during a ride.

The researcher here was setting things up to only get feedback on negative experiences of fairground rides. She needed to establish first what types of emotional responses people had experienced on fairground rides (assuming they all had experienced rides). If she had used such an ambiguous question in an online questionnaire, it might have been skipped by those people (who had never been afraid), or perhaps answered dishonestly (by those people who recalled being a little afraid, but wanted to give the answer they thought was being looked for).

How strongly you agree or disagree with the following statements? Please put a X in the box 1 to 5, where 1 is the most strongly agree and 5 is the most strongly disagree.

	Most strongly agree	Strongly agree	Agree	Neither agree nor disagree	Disagree	Strongly disagree	Most strongly disagree	N/A
Train travel is the cheapest form for most journeys I make								
I choose to take the train rather than taxi/car because of its lower impact on the environment								

Fig. 7.6 Example of a strength of agreement/disagreement with statements

A common form of question used in surveys for people's opinions is the type where you rate the strength with which you agree or disagree with a statement (Fig. 7.6). Responses can be collected on a Likert scale (named after the American organization psychologist, Rensis Likert (1903–1981), who started out studying civil engineering and briefly worked for Union Pacific Railroad). Likert proposed a five-point scale, but this is often extended to permit more "shade" or nuance to responses.

A variation on the Likert or preference scale asks respondents to rate their experience of something. The example in Fig. 7.7 asks train users to rate the comfort of their experience. This type of question might be followed up with an open-ended question to ask participants to explain their rating.

These scales are subjective, and often compound a number of factors into one scale. One of the most famous in the aerospace industry is the Cooper-Harper rating scale to categorize aircraft handling characteristics on a scale of 1–10 (1 being the best) (See Fig. 7.3).

Please put a X in the box to rate your experience on the most recent train (tube or mainline) journey you took on a scale of 1 to 5, where 1 is the most comfortable and 5 is the least comfortable. Please put an X below N/A if something is not applicable.

	1	2	3	4	5	N/A
Air quality						
Train seat						
Lighting-levels						
Acoustic environment						

Fig. 7.7 Example of a preference scale

7.4.5 Who Are Your Survey Participants?

You need to give some consideration to who you are going to survey. The justification of your selection is important. Considering your objective, there will be a population that can give you the information you need, but how big is it? You need to decide whether you have to survey them all, or whether a representative sample would do the trick. Like all experiments the sample size must give valid results. So, "what is the ideal sample size?" Setting aside for a moment the statistical technique for establishing numbers of data points and thus sample size, the answer to this is subjective, and (pardon the implied joke) it is not always size that matters.

To generalize to the wider population, you need to ensure that your sample contains representatives from all corners of the population. The automotive industry was once famously panned for only crash-testing vehicles with dummies that represented the 95-percentile, male population. The problems that arise from not including a diverse population in your research are well-explained in Caroline Criado Perez book, *Invisible Women*.[1]

Use your contacts—this doesn't mean just friends and family, who often may not fit the profile you seek. You have selected a project for your interest, so make sure you meet others who share your interest. A student-group project was involved in a design-make-and-evaluate project of a prosthetic limb adapted for diving. Through the local scuba diving club, they enlisted several experienced divers to help develop their model simulating divers' movement. The diving community were very pleased to help with filming their diving techniques, as they were all enthusiastic for more people (whatever their ability or disability) to enjoy the sport.

7.4.6 The Survey Instrument

The method used to collect information from people is known as an *instrument*. This might be an online questionnaire, an old-school paper-and-pen exercise, interviews, focus groups, or observation.

Survey design is iterative, and may not initially be optimal, but you aim to make it so. Start by identifying the variables that you are going to investigate, and hence what input variables you need to obtain through the survey method so that participants are clear about what information they are being asked for, be clear about what units of measure you are interested in. For example, if you are creating an online form to collect information about shoe sizes of personal protective work wear, make sure you clearly indicate which country's sizing system you are using—or make it a mandatory question that the respondent have to fill-in.

---

[1] Criado Perez, C., *Invisible Women*, Chatto and Windus (2019).

7.4.7 Questionnaires

Questionnaires are relatively easy means of collecting information from a large number of people. Ensure that every response will be used in your analysis. All questions must serve some purpose. Your questionnaire may have branches, so that if the answer to one question is "option (a)", for example, then one line of questioning is pursued, otherwise a different line is followed. "Yes/no" type questions are the easiest to branch, but it is possible to do this for more than two options.

Online questionnaires, such as Formstack or SurveyMonkey, are convenient. They can handle a wide variety of question types, and many of these apps make branched-questions appear seamless. Unfortunately, the free versions are usually limited in their features (or the number of questions you can include). So, before designing your form, make sure you know what format the output is going to be. Output into a spreadsheet permits easier cross-referencing and graphs (what you do with qualitative data will be considered in more detail in Sect. 9.2). Also, consider how the data is stored—will you always have access to it, or only for as long as you pay a subscription?

Consider in the design of your questionnaire how to test for its authenticity. You might use something like CAPTCHA, to avoid the robotic response throwing your data. Asking for a respondent's name is not always appropriate—and it's amazing how much research Mickey Mouse has contributed to over the years! Usually, you have to trust participants to provide genuine information. If your responses are anonymous, then they are perhaps more likely to be candid.

Even in the age of the app, there is a place for the paper questionnaire form, filled in by hand and subsequently easy to destroy. This is much harder to guarantee with electronic means.

7.4.8 Interviews

We'll define an interview here as a one-to-one encounter, either in person or electronically. Some researchers do use group interviews, but the response of one person can influence the opinions and responses of another, so we will take these as focus groups and deal with them later.

There are three types of interview: structured, semi-structured and unstructured. The structured interview is the simplest to design and administer—you have a fixed list of questions, from which you do not deviate, and you know approximately how long it will take to get answers to them. Sticking to a list of questions makes coding the responses for analysis straightforward, and, again predictable for the time and effort involved. So, whilst there are compelling arguments as to why structured interviews are a "good thing", the obvious question is "why bother?" Information collected in a structured interview probably could have been collected as effectively by an online questionnaire.

Unstructured interviews can be more difficult, although they can often provide candid and open answers, if you can just get the conversation to flow. Interviewing people on their territory, or place of work, can be good for obtaining opinions about contentious topics. If carrying out several interviews, use "touch-points", and have a checklist of points you want to cover. If your interviewee sees that you are talking about the same subjects to a number of people or companies, they may be less guarded than if they think you are targeting specific questions to them alone.

The semi-structured interview is the middle-way. Offer interviewees open-ended questions that give them the opportunity to voice their opinions in their own language, but clearly defined questions give a clear idea what information you are seeking. Like the questionnaires that permit you to read through the questions before you attempt to answer them, describing questions before starting permits interviewees to prepare their responses. Dr. Likert (who devised the preference scale) advocated the use of a funnelling technique in interviews so that the conversation might start with very general open questions, but follow an agenda that becomes more specific towards the topic under investigation as the interview progresses. This requires a great deal of design thinking, and anticipation of what direction the initial conversation may take.

Time allocated for the interview is an important consideration. This might be a delicate balance of permitting your participants sufficient time in which to express their opinion to you, and not taking too much of their working day.

Consider how to record interviews. If planning to use a voice-recorder—and there are many apps available to help with this—you then need to decide how you are going to use the recording. We will deal with this in Sect. 9.2; however, one aspect to consider is the duration of interviews—you might decide to transcribe recordings. Be prepared for ten times the length of interview in order to type it up. Voice recognition software might help, but it is not infallible, and still requires your time to check it once converted to text. Similarly, transcription services are available at a price, but still required checking for comprehension, as it easy to mis-hear human speech in an uncontrolled acoustic environment. Often notes, perhaps with a timeline, taken during the interview, can enable more efficient and selective transcription.

7.4.9 Focus Groups

There are some additional psycho-sociological aspects to consider with a roomful of people, being questioned. As an engineer, you have worked as part of a team, and have experienced "group dynamics" even if you have not studied them *per se*. When two or more people work together there tends to be a power gradient, caused by perceptions, self-perceptions, biases (conscious and otherwise), confidence and self-confidence. These can cause interesting results to questions posed in focus groups. For example, a student got permission to survey the occupants of a new open-plan office to find out their thoughts about the indoor environment. Initially, the volunteer representatives (five people of various ages; three women and two men) seemed content to go round the table providing some answers to opening questions. However, as questions focused on various environmental factors, one person dominated discussions, with the others merely nodding or shaking their heads. The student concluded that he would have been better off spending the hour allocated to the focus group on six 10-minute interviews, as he left the group meeting unsure whether the other members genuinely agreed with their highly vocal colleague, or were too cowed to go against the person.

Recording the findings of focus groups, consider using video. This gathers non-verbal communication, as well as the conversation. The problem of coding data gathered by video will be considered in Sect. 9.2.

7.4.10 Product Testing

Product testing combines physical measurement and opinion gathering. The availability of 3D printing has enabled prosthetics to be easily produced, and the obvious corollary of this is testing them, which may well involve the use of human subjects (Fig. 7.8).

This kind of project can use a variety of survey instruments, and record and analyse data in a variety of ways. If approached with the right attitude, there is scope for obtaining test samples and making connections in industry. Don't go out to get "freebies", but you might find you get lucky if you take a scientific approach to comparing the product of Company X with that of Companies Y and Z.

7.4.11 Storing the Data

If your project depends upon survey results, then you need to ensure that they are stored securely. In addition to your need to access them, they must be accessible by authorities within the university who challenge your findings. You need to decide how results of your survey will be kept and how long for. If you have space on your university's server, then you have some assurance of their security. However, your server space is likely to disappear upon your graduation, and termination of your registration as a student.

7.4.12 Testing the Survey Instrument

Part of the design of the survey is about how it is administered for example, how you get a questionnaire to people to fill-in, or how you conduct a product testing. Before

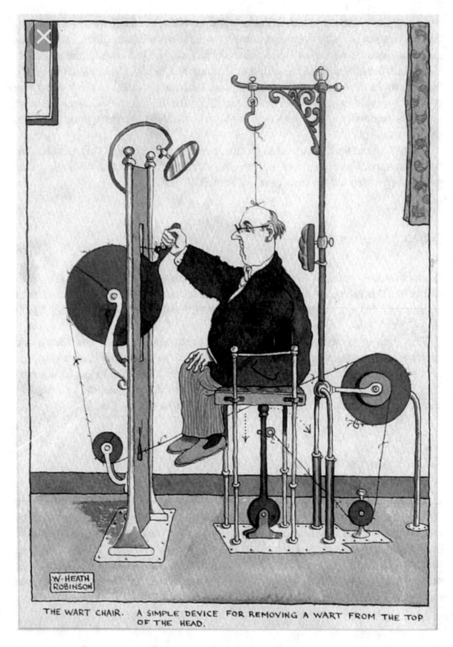

THE WART CHAIR.   A SIMPLE DEVICE FOR REMOVING A WART FROM THE TOP
OF THE HEAD.

Fig. 7.8 Testing a wart removal chair—requiring use of human subjects (by Mr William Heath Robinson)

surveying your whole sample, always select a <u>small</u> subset—or a trusted friend—and run through the survey with them, in the manner of a dress rehearsal. Below is a checklist you could use for a structured interview or questionnaire. Go through the whole process with them, and collect feedback from them at the end. Make any changes to your instrument or the procedure of its administration—and then do it for real!

Survey checklist

- On the "day", provide information to the participant about the reason for doing the questionnaire, and whether they can withdraw at any point
- Explain what will be done with data. Storage? How results will be reported? Attributed? Anonymized?
- Ask and obtain signature on a consent form
- Advise how long it will take to complete the questionnaire (and be realistic), explain things like how many questions there are, and how they are grouped into sections (if it's a long questionnaire)
- If you are carrying out the survey in-person and/or one-to-one, don't let them feel pressured to complete the form, but let them know you are willing to answer any queries they have
- When they've finished the questionnaire, ask them for feedback
- Thank participants.

Chapter 8
Acting Professionally

Abstract In interacting with university and client staff, it is important that students act professionally—this is a valuable practice in preparing for a career in industry or graduate study. This includes generally decorous conduct, appropriate dress, how they behave in meetings and, especially, how they communicate—including how email is used and managing expectations when the project is not going to plan.

© Springer Nature Switzerland AG 2020

P. Gratton and G. Gratton, *Achieving Success with the Engineering Dissertation*,
https://doi.org/10.1007/978-3-030-33192-4_8

8.1 About Acting Professionally

The dissertation is, for most, their first major encounter with the demands of engineering professionalism: a very important skillset as a new engineer progressing their career, whether as an engineer, a researcher or in some allied profession. However, there are unlikely to be any lecture courses in professionalism: many of us pick this up as we go along, or at least learn from being told off for our failures. Here, we've tried to steer you towards best practice in some key areas.

8.2 Meeting Etiquette

Meetings are fundamental to almost all professional practice—engineers regularly meet with each other, supervisors, clients and (when you reach that stage of your career) subordinates. The basic etiquette of meetings varies very little, regardless of the context.

First, it is always vital to agree in advance the time of a meeting and the venue (which often nowadays may be on the phone, by a teleconference facility or application such as *Skype* or *Google Hangout*—but those in themselves become the venue). At the very least there must be a clear understanding of who is attending, the maximum meeting duration and of the subjects to be covered—although more normal and preferable is to have an itemized agenda, agreed in advance, for the meeting.

Agendas do not need to be overlong for dissertation meetings—it may sometimes be appropriate to add timings if anybody's concerned that the meeting might overrun, but there's no need to be excessive about this. For example, the following might work well enough:

Meeting Agenda – 10th February

M. Patel (student), Prof B. Smith (Supervisor), Mrs P. Gratton (client, by phone, first half only)

1030—Lab results so far

1045—Progress on prototype design and build

1100—Latest literature review

1115—Structure of dissertation

1125—Next meeting and AOB

More than this, for a meeting about a dissertation is almost certainly excessive, and even here this may be more than required, but it indicates general best practice.

Beyond this, always remember:

(1) Be punctual.
(2) If you can't make a meeting, or need to change an agenda or timings, make sure everybody knows as soon as you do.

(3) Remember that everybody has their contributions to make, so when it is somebody else's turn to make their point, keep quiet and let them.

(4) Make sure it's clear whose job it is to make and distribute notes ("the secretary"). It's likely that in dissertation project meetings—this will be the student. Such notes or minutes can usually, however, be very brief, and limited to what was discussed and on any decisions (particularly "actions"—where somebody is tasked to do something.)

(5) Just to repeat: be punctual.

In both business and academia, the use of electronic meeting notices is normal nowadays—any email system and many other forms of "business efficiency" software include this provision. Unless specifically objected to, these are likely to be acceptable to everybody and often preferred—but, there is seldom harm in asking what people's preferences are.

8.3 Dress Codes

Okay, we get it—you're an engineering student. You live most likely in sweat pants or jeans, trainers and a hoodie—or the T-shirt from the last gig you attended. That's fine; absolutely nothing should stop you being who you are in your normal cycle of lectures, labs and meetings with your supervisor and fellow students. Equally, however, most engineering workplaces do have a written or unwritten dress code of some sort, and you should where appropriate respect that. What there won't usually be is a universal rule of best practice, so you might need to do a bit of detective work.

In labs and workshops, mostly it's just "sensible clothes"—long trousers, nothing hanging off you that can snag, closed shoes. Sometimes you'll need to wear a labcoat or overalls if the work demands it. This, of course, hasn't changed since you were at school.

In meetings, presentations or visits to company premises, very often how you dress can affect how you are perceived. Of course, anything you're wearing should be clean, free of holes and rips, and probably not covered in logos (UNLESS they are the name and/or logo of your engineering project team, which often can carry a lot of weight). After that, it's easiest to just ask, most likely your supervisor (Fig. 8.1) or client if you have one. If in doubt, a business suit seldom goes wrong; but, it may also be unnecessary—nowadays smart casual trousers or skirt, and a polo shirt may well be absolutely fine. (Indeed, at NASA that combination seems to be compulsory these days!) Avoid sportswear: it is seldom seen as professional in any engineering or science environment.

All of these rules about dress codes will probably serve you well later as you enter the workforce and, especially, for job interviews (where the business suit fall-back becomes particularly sensible). But, universally it is the case that the way

Fig. 8.1 Whilst hopefully
you can take useful dress
code advice from your
supervisor, there may be
exceptions (portrait of
Professor Theophilus
Branestawm by Mr William
Heath Robinson)

you present yourself visually will influence how fellow—particularly senior—professionals regard you. Whilst you might wish it otherwise, and think that only the quality of your work should matter, the reality is different.

8.4 Phone and Electronic Messaging

Once you have got to know the people you'll be working with, you'll establish what acceptable practices are for communication, but starting out with some good practices is wise.

Nowadays emails (or equivalent electronic messages, e.g., via the university's online teaching systems) are the most common way to communicate with colleagues, supervisors and clients. They're immediate, everyone can send and receive them at their convenience, and include attachments, pictures and so on. However, it's still very possible to get them wrong.

Bombarding somebody with daily messages about every small issues that you think matters is likely to annoy, and most such emails will end up not getting read. So, think hard about what's appropriate and necessary—and take the time to distil that down to the key message that genuinely needs to be read. Communicate when needed, and only include the information that's needed—not more often, and no more. Take care over your spelling, grammar and syntax, and avoid "text speak", as this will really undermine the perception of you as a professional engineer.

Most messaging systems, mailtools and so on, include a system for a message footer. **Use it**! Busy people may well know many people called Mike, or Sue, or Mohammed—so a simple first name at the bottom of a message is seldom acceptable. As an absolute minimum, a full name, preferred name and role should be there, for example:

Mo

Mohammed Abadi
3rd year Student, BEng Civil Engineering

Or

(Dr) B M (Barbara) Rosenberg
Associate Professor, Department of Electronics

Of course, other information can be added—phone number(s), twitter handles, LinkedIn URL and so on. Regarding use of the telephone, nowadays it's unusual to routinely telephone colleagues, and less so supervisors, especially on their mobile (cell) phone. So ensure you know what any person's preferred etiquette is, and don't be afraid to politely make it known your own preferences also (Fig. 8.2).

If using personal email, or any other communication method where you choose your own username, take some care with this. Anything that's just a shortened version of your own name is likely to be fine: but avoid anything humorous or suggestive. We really have seen students using variations on "BigBoyBob@domainname.com" or "SexySarah@sillydomain.net"—just don't do this! Something more meaningful but factual is probably okay—so "JohntheRacingEngineer" or "SitaSpaceScience" are harmless enough if they make you happy, but don't necessarily help your cause as a serious future engineering professional. If in doubt, be boring!

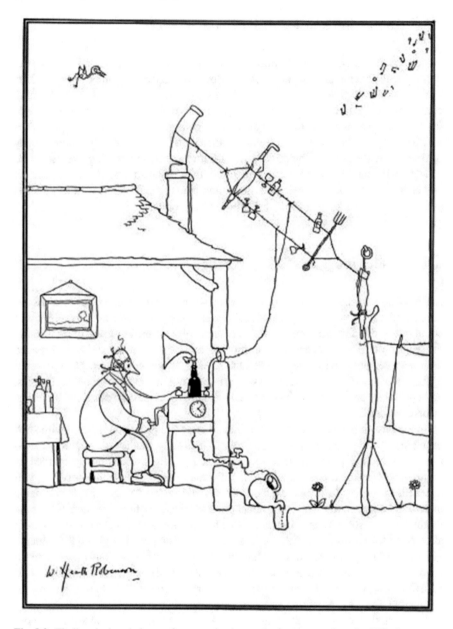

Fig. 8.2 Finding the best balance of communication methods can sometimes be difficult

8.5 Monitoring and Control of Your Project

Project management is mostly about monitoring the progress against the project plan—the utilization of time, budget and resources. Hence, recording your activities against what you planned, for example in your Gantt chart, is vital to maintaining control of your project, and must be a regular feature in meetings with your supervisor (and client). If you notice that any element is going to deviate from the plan, then it is better to re-plan, and avoid failure, rather than having to ask for more money or time at the deadline of the project. In earlier chapters on planning, we discussed *run-offs*—if you do have to use one of these, then make sure to check what effect any changes have on your risk assessment and ethical approval, as well as finances and time. A controlled change is unlikely to be a problem.

8.6 When Things are Going Wrong

It would be surprising if nothing ever went wrong during the course of a dissertation project. On a purely professional level, equipment can break, results don't come out in the direction anticipated, a design solution proves to be impracticable or key people (client, supervisor, an important stakeholder…) may have to reduce their involvement due to unrelated demands. As the student, you may have personal problems yourself—mental health issues are sadly not uncommon amongst students, and absolutely anybody may have to deal with money issues, physical health issues or bereavement.

In the past, it has been normal, and indeed regarded as healthy to deal with such problems, "soldier on" and just do your best. Thankfully, that is no longer the case—indeed concealing problems that may prevent a project succeeding is regarded as highly unprofessional. Universities also universally have counselling services, you should have access to a medical practitioner, and in most university departments there will be somebody who is a contact point (typically with a title such as "course mother" or "personal tutor") for students with problems, and whilst a supervisor isn't allowed to stop a student from failing, they are also supposed to give their students every opportunity to succeed. In all likelihood, there's also a mechanism, if there's a need to use it, to make concessions in terms of marks and deadlines for a student who really is being prevented from delivering by circumstances outside of their control (typically known as something like "extenuating circumstances").

So, when things are going wrong, TALK ABOUT IT. Identify who needs to know (when problems are personal, this should be a very short list; on the other hand, if it's an equipment breakdown clearly everybody involved needs to know, promptly and clearly: ideally including a plan to resolve it). Do so promptly, and to the right people, with enough information so that together the problems can be resolved. If you do take the old-fashioned approach of concealing the issues and trying to persevere through them—the most likely outcome is unlikely to be one you'll enjoy.

And don't panic. Even if all of the equipment fails, the client withdraws from the project and your supervisor is unavailable for weeks at a time dealing with a family emergency, it is still entirely possible to get top marks in your dissertation. The experience may have been much less satisfying than hoped for, but if the writing, background research, problem-solving, analysis and conclusions are all good enough, excellent outcomes are still possible, not to mention quite probably a very strong learning experience. That said, we hope you do have a much better experience than that—and the vast majority of students do.

8.7 Managing Discontent

Disagreements will occur. A student may be unhappy with the way their supervisor has marked something, or any person in the team behind a dissertation may feel that their position isn't being respected, or that somebody else isn't communicating adequately.

Good advice in such circumstances is, wherever possible, to deal with this personally and ideally face-to-face—if not then at the very least on the telephone. Remember that absolutely nobody responds well to public accusations or having to deal with some external person to the team investigating their conduct, and at the very least such an event takes people's valuable time away from the task in hand. Students do unfortunately sometimes forget in particular both that their supervisor's judgment is likely to be considerably sounder than that of a student (and sometimes, also that of the client) but also that a university will inevitably back up their staff's academic judgment unless there is glaring evidence of malpractice. Creating conflict over a disagreement will only damage a working relationship that potentially could become important again many years later (both the authors of this book are in professional contact still with their dissertation supervisors, many decades later—it's a small world). If, on reflection, you do find that you have stepped out of line, you will lose little face by apologizing for having done so; conversely, difficult students' (or for that matter academics' or engineering professionals') reputations can follow them, damaging their careers, for many years later, even over apparently relatively trivial incidents very early in their career.

Chapter 9
Presenting Data

Abstract All engineering projects create data which, may be broadly divided into two forms: quantitative and qualitative. It is important that all data is obtained and presented in a form that allows it to be well understood, without the report becoming nothing but a bulky presentation of data alone with no analysis. Quantitative (or numeric) data may be presented in tabular or graphical form, normally accompanied by explanatory metadata. Sketches, photographs and drawings are grouped here with quantitative data and guidance given on their preparation and presentation. Qualitative (descriptive) data is also important for engineering—for example in evaluating stakeholder opinions or user experiences. Examples are given of how this might be obtained and presented, as well as integration with numeric data.

© Springer Nature Switzerland AG 2020

P. Gratton and G. Gratton, *Achieving Success with the Engineering Dissertation*,
https://doi.org/10.1007/978-3-030-33192-4_9

9.1 Quantitative Data and Illustrations

9.1.1 Introduction

In writing an engineering dissertation, you are presenting your first large piece of independent work as a future professional engineer, or engineering researcher. It is vital in that work to be able to present "hard" qualitative data and use it well within your description and arguments. It's important to appreciate that this <u>does not</u> mean simply filling your dissertation up with lots of tables and graphs—on the contrary simply packing a report with large amounts of data can potentially leave the reader very unimpressed and result in poor grades. All figures, data and illustrations must be there as part of the *narrative flow* of the report—they must be part of a flow of information that leads to the eventual conclusions and recommendations. It may well be that at the drafting stage of your report you will want to include an excess of information, to be trimmed back in the final submission; if you do, that's fine, but don't forget to go back and do that trimming.

A further requirement of all data that you present, is that it must be complete and self-contained. Consider that every figure in your dissertation may potentially be copied and used in isolation; so it's vital that it is presented clearly enough that it can be used in that way, and makes unambiguous sense. Consider Table 9.1. This is poor, and is not what should be presented in a well written dissertation. Climb performance of what? What are the airspeed units? What does the time represent? Was the climb rate calculated or displayed? On what occasion were the data obtained. There are many things that can be misinterpreted in this table, simply because of the lack of clarity.

,Compare this to the modified version in Table 9.2. The numerical data is the same in both cases, but the second example contains significantly greater metadata—that is "data about data". The heading, three lines of information about test conditions, and column headings concerning data units and source all fall into the category of metadata. An additional term that can often be attached to a set of data which applies to the conditions of everything in a given table or figure is common data.

Table 9.1 Minimalist presentation of tabular data

Climb performance			
Indicated airspeed	Calibrated airspeed	Time	Climb rate
35	38	44.5	675
40	42	40	750
42.5	43	38	790
45	44	37	810
47.5	46	41	730
50	48	40	750

Table 9.2 A more complete presentation of tabular data

Cessna 150F climb performance

Climbs were carried out from 1250 ft to 1750 ft QNE on 23 September 2018. Loading was as per 75 kg front crew, no ballast or observers, 45 l fuel (1,045 lb, 10.7"CG). The following results were obtained

Indicated airspeed (mph)	Calibrated airspeed (knots)	Time to climb 500 ft (s)	Calculated rate of climb (feet/min)
35	38	44.5	675
40	42	40	750
42.5	43	38	790
45	44	37	810
47.5	46	41	730
50	48	40	750

Check your institution's guidelines on labelling figures. This book for example has consistent figure labelling in line with the with the publisher's norms for textbooks—not necessarily the same way as your dissertation needs to be, but with the same requirement to be clear and internally consistent. If the guidelines don't contain anything, it's probably legitimate to make up your own (sensible!) standard or just use your word processor's norm—but if in doubt, check with your supervisor, or the module leader for the dissertation. The same is true of the location of illustrations. Historically (before we all had word processors), it was normal to have a typed dissertation with no embedded illustrations, and for all of the figures and tables to be in separate pages at the end. Now, it is more normal for figures and tables to be included with the text—but some institutions will use writing guidelines that still adopt the old fashioned approach. Basically: never go outside your institution's dissertation writing guidelines, and always be consistent within your report document.

9.1.2 Tables, and Numbers Within Text

The tabular presentation of data is an essential part of most scientific and engineering reports, and the dissertation is no exception—whilst the earlier cautions about not simply filling up the report with model outputs or experiment results remain equally important.

Tables can of course contain more than just numbers—and some numbers may not really be numbers (see the end of this section on numerical expressions of opinion), but there are some universal issues of best practice, whenever information is presented in a tabular and/or numeric format:

– Avoid inappropriate (which usually means excessive) precision: for example expressing that the length of a component was measured at 23.0571 mm, or air pressure was 998.456 hPa. In both cases it's highly unlikely that measurements genuinely could be taken to an accuracy of 6 significant figures; in both cases 4 significant figures: 23.06 mm and 998.5 hPa are probably as accurate as can reasonably be determined, and often only 2 or 3 significant figures are defensible, depending upon the error analysis. Always ensure that the precision used is appropriate to what's being described. [In terminology, *precision* defines the resolution with which values are described, whilst *accuracy* defines the maximum error between the error returned, and (the probably unknown) reality.]

- Simple tables are poor for expressing data in more than 2 dimensions—consider using shapes, colours or graphical representations such as carpet plots when it gets more complex than that.
- DON'T fill your report up with tables—only in the main body of your dissertation include enough tabular data to permit the reader to follow your arguments. If you really *must* include further tabular data than is strictly essential, place it in an appendix—but for a dissertation it is highly likely that you will have much more data than it's wise to include, even there. Mention what data exists, but only include what's absolutely essential for the reader to understand your work. [In more complex reports such as journal papers, or PhD theses, it's increasingly the case that such additional information is placed on a non-expiring webpage for access by the inevitably very small number of users who *really* need to access it.]
- Be consistent in the units you use. Different countries, universities and engineering disciplines will have different norms—and some of these will not even be standard SI or Imperial (termed often in the USA "British Units", despite the fact that the British mostly stopped using them in the 1970s)—for example that aeroplanes and ships usually operate in knots and nautical miles, road vehicles in mph or kph, and much electronic engineering still uses centimetre-gramme-second or cgs units, only the last of which is strictly SI. The use of non-scientific units is not inherently bad, but regularly switching between units is always bad: BE CONSISTENT, and of course, be clear. It may often be worthwhile explaining in multiple units, but again consistency is essential—for example don't describe a distance as 1.1 nm (2,037 m) in one place, then later as 2.04 km (6,683 ft)—this is highly confusing to the reader, who may conclude that you are trying to conceal something from them; at best they'll be annoyed about this—and since the reader is probably also marking it, that won't end well.
- Again, be consistent in presentation of tables. If your institution has clear guidelines on how they want tables presented—use those guidelines. If it does not, pick a standard and stick to it and you are unlikely to be criticised. Your word processor is likely to have a wide range of choices, most of which are acceptable—but there is seldom any justification to using more than one format in a single document.
- Where something has the same value for every item in a table column or row, consider just taking that out and including it in the metadata/common data.
- A note on international norms. In Britain and America, it's normal to describe numbers with thousands separated with commas, and the decimal place separated by a dot: so 123,456.789 would be a normal representation of a "six digit" number expressed to three decimal places. In France and Germany the symbols would be

Table 9.3 Illustration of a poorly constructed data table

Observations of instrument performance in August

Time	Voltage (V)	Current (mA)	Horizontal movement	Vertical movement	Indicator LED colour
0:00	12	1	1"	5 mm	Red
0:30	12.4	0.95	1.1"	4 mm	Red
1:00	12.55		1.15"	2 mm	Red
1:30	13.1	0.916	1.3"	−1.5 mm	Red
2:00	13.05	0.92	1.255"	−0.2 cm	Red

reversed so this becomes 123.456,789. In India, Pakistan and Bangladesh conversely: lakh (1×10^5) is a common concept so 1,23,456.789 will be more normal [All of these are of course the same number—one hundred and twenty three thousand four hundred and fifty six, decimal point seven eight nine]. None of these are incorrect in the right place, but using more than one format in the same document, or leaving the reader unclear about the format in use, is always incorrect.

Table 9.3 is provided here as an example of how not to present tabulated data. As an exercise for the reader, see how many examples of poor practice you can find within the table, and separately re-draw it as you might want to read it, if you were marking a dissertation. Once you have re-drawn it, compare your version to Table 9.5 at the end of this section.

9.1.3   Graphical Data Presentation

All engineers and scientists are of course trained to draw graphs, both computer generated and hand-drawn. Your university will have guidelines as to what is acceptable—some may prefer computer generated graphs (or polars), whilst some others may permit or even encourage hand-drawn graphs. This is one of many things that the best students will have familiarized themselves as part of their pre-dissertation work-up. As an example of good practice consider Fig. 9.1.

Don't worry if this graph is about something significantly out of your area of expertise; the important thing here is to see the major points as to how this graph is presented. Specifically:

– Common data (or metadata) is clearly presented above the graph.
– All axes are labelled with both what they represent, and the units.
– There are error bars on each data point (we'll assume for now that the calculation of the sizes of those error bars are based upon calculations also shown somewhere in the report and which bear scrutiny).
– Curve fits go through the error bars and whilst they interpolate between data points, do not extrapolate beyond them. (Often it may be required that the curves

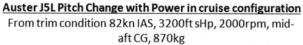

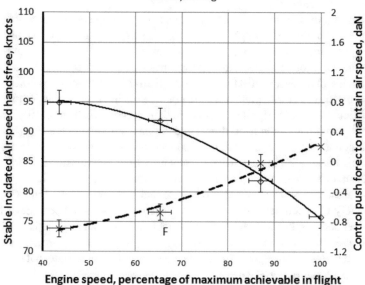

Fig. 9.1 Graphical illustration of some flight test data

also follow a shape which corresponds to applicable engineering theory, this of course is a very discipline and sub-discipline dependent requirement.)

9.1.4   Consideration of Errors

An important part of both real scientific analysis, and academic work, is to consider the source and magnitude of errors in your data. All quantitative data contains, or has potential to contain, errors—whether this is errors in model outputs due to assumptions that are made about the processes and structures being modelled, instrumentation (accuracy, resolution and calibration) errors in observed quantities, or errors in the value of derived quantities.

How this should be expressed will depend massively upon your engineering discipline, and the dissertation sub-discipline: whether the project is analytical or experimental, and how your data is being expressed. Sometimes it may be a simple case of numerically combining data and presenting it graphically as error bars, blocks or ovals; sometimes it may be more descriptive. The best approach with presentation of errors is likely to research and draft an approach, then discuss this with your supervisor.

9.1.5   Sketches, Schematics and Illustrations

Sketches and Schematics

It is important to appreciate that a *sketch* in an engineering context does not usually imply the sort of thing that might be generated in a primary school, or on the back of an envelope during a discussion. Engineering sketches are illustrations that are used to illustrate ideas or concepts, and require considerable work to ensure that they are clear and unambiguous. Whilst they are unlikely to contain dimensions, tolerances or detailed materials information—they are nonetheless important tools for presenting information and require significant effort in their preparation. Historically, sketches probably would have been hand-drawn and engineers trained in the skill, but in the modern age are far more likely to be generated using a piece of drawing software, and may well be a 2-dimensional representation of a 3-dimensional form, or of a mechanism or mode of system operation. If in doubt check your university's guidelines—but the same high standard will be essential, whatever the medium. Guideline are unlikely to specifically state views on *whether* you should reproduce figures from other sources, but wherever you reasonably can—avoid it: there is much more credit to creating and using your own illustrations, whatever they are of. Of course, where you do reproduce figures from other sources, it is vital to state that source (for example as shown in Fig. 9.2).

Figure 9.2 shows the side view of an aircraft safety harness. This would be classed a as a sketch or illustration, as - not being to scale, drawn to a defined technical drawing standard, or containing titling and revision information—would not be described as a drawing.

The use of colour can also be enhancing in sketches, and is usually easy to achieve now with either electronic submission, or the use of a colour printer. For example

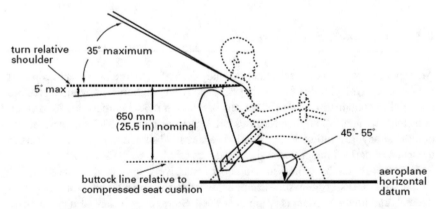

Fig. 9.2 Sketch of the side view of a safety harness [From EASA Certification Standard CS.23, Light Aeroplanes]

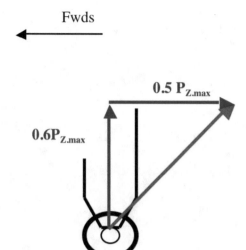

Fig. 9.3 Illustration of loads acting on a wheel

Fig. 9.3 shows the forces acting on a wheel. This could be all drawn in black on white and would just about be readable, but the simple separation of the black structure, blue defined forces, and red resultant force makes this much easier to read and explain.

Photography

Photographs also belong in a dissertation (or most other forms of engineering report)—they serve often to provide easy documentary evidence of items being discussed. Virtually everybody now owns a good quality digital camera, if only one built into a smartphone, so it's possible to take photographs easily, and then paste them into word processor documents. However, as with all other things, quality matters and a little effort can be used to massively improve their usefulness. For example, look at Fig. 9.4 showing an air data instrument. You can see it, and with some description, it can be seen and the main parts identified—but equally the item of greatest interest is small within the field of view, and a lot of background clutter makes this hard to make out the details.

Conversely, Fig. 9.5 shows the same item, but much closer, and by use of flash and a large aperture (giving reasonably shallow depth of view) the item of interest appears lighter and more distinct than the distant background. Thus, it is far easier for the reader to make out the details of it. This can of course be further enhanced by clear labelling, again easy to do in software.

Much more control is available when, for example, photographing items in the workshop or lab, when it may be possible to put a piece of cloth or large sheet of

Fig. 9.4 Air data sensor on an X-31 research aeroplane

Fig. 9.5 Close up of the X-31 air data sensor

paper behind and under the item being photographed, effectively eliminating the background altogether. Other useful hints are shown below:

Scale   If the scale of an item is not obvious, place something of known
                       dimensions (for example a ruler or large denomination coin) in shot
                       and adjacent to the artefact being photographed

Colour contrast	Try to use a dark background when illustrating light coloured objects, or a light background when illustrating dark coloured objects
Focus contrast	If aperture is controllable, use of a large aperture (small "f" number, e.g. f1.4) produces a short field of focus—so anything out of range in the foreground or background may hopefully be blurred and out of focus, emphasising the important photographic subject [f-number is defined as the ratio of focal length to aperture diameter of the imaging device.]
ISO	Where these are adjustable, you should aim to use a low 'ISO' number (equivalent film speed) to minimise grain in the image. This may necessitate long exposure times, which can require the camera to rest or be mounted onto something so as to avoid blurring.
Sharpness	In poor light, steady the camera on something, or use a tripod. Using a timer (say, 10 s) can give the camera an opportunity to stabilise on its support before it takes the picture
Detail	Try to have the main light source diffuse, or behind the photographer—avoid silhouettes and shadows that make it hard to make out the detail of the item in the photograph

For example, Fig. 9.6 shows a component in a way that might be included in a report. This has been held in space with a large aperture/narrow depth of focus, making the detail extremely clear with the background out of focus to the point of being indistinguishable. The fact that it is being held in the photographer's hand provides a clear visual sense of size, and it is sufficiently darker than the background to be clearly visually separate. [Mechanical engineers will probably also recognise that the illustration clearly shows the inadvisability of using peening to hold in a bearing, and a classic mechanism for fatigue crack initiation.]

Fig. 9.6 Photograph of a failed engineering component

9.1.6   Drawings

A drawing, in an engineering context is something that is typically used to provide detailed manufacturing or construction instructions. It will invariably be prepared and presented in accordance with strictly defined drawing standards. As with sketches (see earlier), these may have historically been hand-drawn, and a student displaying the right skills is unlikely to be penalised for presenting hand drawings, but the use of CAD (Computer Aided Design) is far more likely. Standards will vary between industry sectors (Table 9.4), and students in any engineering discipline will hopefully have been familiarised earlier in their course with the standards in their discipline— use the right standard for what you're doing.

When presenting drawings, like anything else, you must consider how these fit into the overall scope of your dissertation. The dissertation is unlikely to be an exercise in technical drawing in itself, and as with other data, examiners will not be impressed if the dissertation report appears to be padded out with drawings. Including them may often be appropriate, particularly if the project has been a design-and-build one, but it's likely to be most appropriate that they are there to illustrate the progression of

Table 9.4   Some of the main engineering drawing standards in present use

Mechanical (including aerospace and marine) engineering norms

UK: BS 8888:2017 Technical product documentation and specification

USA: ASME Y14.5-2009 Dimensioning and Tolerancing: Engineering drawing and related documentation practices

NASA/TP–2015-218755, ENGINEERING DRAWING PRACTICES, AEROSPACE AND GROUND SUPPORT EQUIPMENT, May 2015

Other international: ISO 8015:2011(en), Geometrical product specifications (GPS)—Fundamentals—Concepts, principles and rules

From the perspective of a project student, for the vast majority of projects, the subtle differences between these standards, and between these latest versions and earlier versions are small and likely to be of trivial importance. Most engineering drawing or technical drawing textbooks, and there are many good books available, will present a relatively generic set of best practices that if adhered to will be adequate for almost all purposes, and be close enough to compliance with all of the latest formal standards as to be unproblematic. It is far more important that drawings are clear, well presented, and include fundamental information such as titling, dimensions, tolerances and version numbers.

Electrical & Electronic engineering norms

Few "paper" standards of symbology still exist, those which do are essentially derivative of the IEC online electrical symbols library at http://std.iec.ch/

Building services norms

UK: BSRIA Building application guide series/doc. No. BG6/2014: Design framework for building services, 4th edition

Railway engineering norms

UK: BS 376-2:2015, Railway signalling symbols. Specification for symbols for circuit diagrams

Maritime engineering norms

ISO 47.020.01—General standards related to shipbuilding and marine structures

Table 9.5 Corrected version of Table 9.3

Measured instrument performance: external flamwidget
Common data
Date: 23 August 2018
Design standard: As per drawing 12 (general assembly) V2.3a
LED indicator red in all tests

Time from experiment start (mm:ss)	Potential difference (Volts)	Current (mA)	Horizontal movement (mm)	Vertical movement (mm) [+Ve extended]
0:00	12.00	1.00	25.4	5.0
0:30	12.40	0.95	27.9	4.0
1:00	12.55	Not recorded	29.2	2.0
1:30	13.10	0.92	33.0	−1.5
2:00	13.05	0.92	31.9	−2.0

thinking, rather than to demonstrate your ability to produce a pack of drawings. (Well executed manufacturing drawings are most likely to lead to well-made components, and some universities may ask you to separately submit drawings as part of the dissertation's supporting material.)

9.1.7 Solutions to Problem

See Table 9.5.

9.2 Qualitative Data

9.2.1 Types and Sources of Qualitative Data

Qualitative research is about extracting meaning and identifying concepts—rather than measuring occurrences. All too often, when one encounters qualitative data in engineering reports, the presentation and analysis stop at counting. Whilst this might be justified as frequency being an indication of strength of feeling, this is not the only kind of analysis that can be done on qualitative data.

Data may be textual responses to questions, narrative observation of how people use something, or how they behave in specific circumstances. This kind of observation is often done, and recorded, via discreet video. For example, a great deal of observation was used to establish how people use the ticketing and automatic barrier-entry systems for the London Underground system, so that a system that is as free-flowing as possible could be created. The engineering data used to solve the bottle-neck caused by ticket checking was compound: the flow of people was both observed and modelled using computational fluid dynamic algorithms to establish the optimum distances between disembarkation and barrier, and ticket-purchase and barrier. In addition to considering the physical environment, the means of paying for rail travel was dissected into its component parts, and systems of getting payment devised to include a diverse population of travelers.

9.2.2 Logic—Foundation of Analysis

Before considering how to present the qualitative data and forms of analysis with which to process it, we need to think about logic. The kind of logic that most of us grow up considering as defining logic is *deductive logic*. A number of statements, or premises, are considered in a process of reasoning, subject to conditions and links, and a conclusion is reached that is certain. For example, in an instance, all the lights go off and electrical equipment stops working; lights and electrical equipment depend upon electrical connection; therefore, if they all go off at the same time, there has been a break in the electrical connection. All the evidence stacks up and supports the conclusion (of a power-cut).

There are times when things are less certain. This is when we use *inductive reasoning* to come up with a theory. For example, time-lapse photography is used in an open-plan office to survey movement. The experiment records (at 60 second intervals) that people are seen standing in various areas of the office on at least every sixth frame during the working day. In theory, according to inductive logic, the lights in the open-plan office can be switched off after seven minutes, as the occupants are not still for more than six minutes. So the passive infrared detector (PIR) switch is set with a seven minute delay for "off". Now, I'm sure you can see the problem with this, but it is the logic that thousands of asset managers all over the world use to save on their electricity bills due to no-one taking responsibility to turn off the lights at the end of the working day. There is clear evidence to support a conclusion, and the conclusion is probably true under all circumstances, all of the time, but it is a little less certain. When your analysis depends upon observations, you might not be aware of all the possible phenomena to be observed. The number of observations matters in so far as it helps you put a figure on the probability. So, in 72 hours of time-lapse (4,320 frames), the longest count between frames when there was movement in the office was 720, the second-longest count was 80 frames, followed by 52 frames, then the next longest was 6 frames, followed by 4, 3, 2 and 1. There were no intervals of 7–51 frames in duration. So, the most time anyone is stationary is probably six minutes.

The third form of logic we need to consider is known as *abductive reasoning*. This is most relevant to the kinds of qualitative data analysis you are likely to encounter in engineering projects. Based upon what has been observed, logical inference is drawn, and these lead to a plausible conclusion. Again, this is the "most likely" conclusion, and contains some uncertainty. Thinking about the time-lapse in the example above, who or what was the cause of the movement after a long period of no movement? Was it the janitor or was it a ghost? The "best available" judgement in the given circumstances is that it was someone cleaning the office. The abductive conclusion—the most plausible explanation—is that it was the janitor—yikes! Have a Scooby-snack!

9.2.3 Presenting Qualitative Data

Everything that was said in the introduction of Sect. 9.1 about presenting quantitative data holds true for dealing with qualitative data also. Likert scales to reduce expressions of opinion to a point on a five-point scale or a ten-point scale (like Cooper-Harper) are an attempt to quantify qualitative properties, but don't kid yourself that turning an emotion into a number (see later in this section), is the ultimate way of dealing with that emotion. The validity of calculating mean and standard deviations was discussed in the previous section. You might consider establishing the median of your data-set, if you have sufficient points. From the time-lapse example above we have 8 points: 1, 2, 3, 4, 6, 52, 80, 720. The median (middle) is between 4 and 6. Calculating the mean of these two, we establish that the median is 5. The mode is the most frequently recorded value. Apparently, that was 4 frames in the example above. These values might be interesting, but what do they tell you? It's a busy office with lots of movement. It tells you nothing about why people are moving, nor who they are. It doesn't provide you with any more information from which you could make further suggestions for energy savings (and because it doesn't capture the distances they travel, you cannot reasonably suggest capturing all the kinetic energy exerted on the floor to convert into electricity for the lighting). Therefore, the limited quantitative data available is totally inadequate for understanding the situation.

9.2.4 Presenting Interviews

Collecting information through the use of interviews is a great opportunity to capture some of the emotion of an investigation—as well as the facts of the matter.

This was mentioned in Sect. 7.4. If you decide to have a whole interview transcribed as part of your analysis process, you have the option to include the full transcription, although it is recommended that this is something for the appendix rather than the body of the report. Table 9.6 contains a brief extract of a fictitious interview undertaken as part of the design development of the Mars rover. Note how little things have been added to convey the pace of the interview. The three dots [...] are known as ellipsis. These are generally used when something has been omitted. It is a useful convention in interview transcription when you can't hear a word clearly (and don't want to make a guess).

Table 9.6 Example of the presentation of an interview transcript (with acknowledgements to Andy Weir and appreciation for his book "The Martian")

Start time of segment	Speaker	Discourse Extract from an interview with Mark Watney on the UX of the Mars rover Int = the interviewer MW = Mark Watney	Thematic markers
11:13	MW	You have to realise it was never designed for a journey. Three klicks is like a walk in the park on earth… heck, that's from here to the railway station… you could easily walk it! With your suitcase, well, maybe a roll-along. And I took it on a three THOUSAND [raises voice] klick drive	
	Int	So, we've discussed the need for additional electrical power. Can I ask you now about the cabin? Can you tell me about the cabin environment?	
	MW	Well, it got pretty fetid… and there's not much room to get out of a suit…	
	Int	Is there a protocol…	
	MW	Well, obviously, no-one ever thought the rover would have to double-up as a changing room, but just like a prom dress, I wanted to keep the suit special for when I really need to make a good impression!	
	Int	Mmm, mmm, OK. And what about the seating position? Was it the right position to reach all the controls you needed to reach?	
	MW	Quite the opposite! [laughing and nodding head] It was all too easy to hit something, and inadvertently activate something when I rolled over in my sleep! And there's another thing: Why put those… hard bits between the seats, so that they are incredibly uncomfortable to stretch out over! [pauses for a long time] I don't suppose there is a protocol for sleeping in a rover either!	
	Int	Going back to the cabin environment, how would you describe the aircon system?	
	MW	Inadequate. Next question	
	Int	Would you say that the vents were appropriately placed to get a good air flow on your face? And adequately adjustable, so that you didn't have to have the air on your face?	
	MW	Mmmm? Yes, and yes. I was just glad to have… some breathable stuff that didn't rely on a mask, if you know what I mean	
	Int	So, how would you describe the air quality?	
	MW	Adequate. Did I mention how smelly it was? You've got to love your own stink! You know, it was probably just as well I was on my own in there! [waves hand in front of his face; makes disgusted expression]	

Weir, A., The Martian, Ebury Digital (2014)

Another convention that has been used in this example transcript is the use of square brackets [ ] in the text in which to include additional notes about things observed in the interview—like laughter or gestures recorded by the interviewer or an observer. These all add to the record of human interaction.

It is <u>not</u> essential to include verbatim transcripts—or even voice recordings—although, given how easy it is to make voice recordings, your results might seem less plausible if you don't offer them. If you decide not to make a voice recording, but to just make your own notes of key ideas or statements during the interview, this is acceptable. However, you might still like to include a few pertinent direct quotations to portray the nature of the interview. For example, if you want to convey how crammed the user felt in the cabin of the Martian rover the direct quotation "no-one ever thought the rover would have to double-up as a changing room" helps to get this point across to your reader.

9.2.5 Critical Discourse Analysis

Having recorded human interactions, you are now able to subject your recording to a *critical discourse analysis* (CDA). Discourse generally refers to anything involving talk or text—so the same techniques can be applied to those open-ended questions in your questionnaires. CDA is essentially about language, and how it is used, and its function in human interaction. This is multifaceted and interdisciplinary. If you have studied graphical design, then you have probably encountered *semiotics* in relation to symbols and their interpretation—these too have their role in CDA as an extension of language. There are specialist text books on how to do CDA, and the sociology and anthropology sections of your university's library will carry case-studies that have used these techniques. This description is the cut-down version for engineers.

Identifying the Data in the Discourse
Firstly, evaluate the response rate for your instrument. Of your sample, establish what proportion have responded. There is no correct answer for this, but you will be credited academically for meaningful consideration.

The second issue to be considered is about the quality of participants who have responded. From the qualifying characteristics that you surveyed, are they the most suitable to supply the information and opinion you want? For example, if you are wanting to improve facilities for rail travelers, consider the relative value of responses from (a) the person who commutes for over one-hour every day, and (b) someone who uses the train less than six times a year for journeys of less than half-an-hour, and (c) a third person who hasn't been on a train journey in the last twelve months. If the responses are mainly from type (c) people, you are not going to learn much from their current experience of train travel, but they will have valid aspirations and desires for future experiences.

The third aspect of the discourse analysis should relate back to why you thought it would be a good idea to collect this qualitative data. Return to your objectives.

Review the design of your survey instrument, and within that, where relevant, each question posed. The first stage of this aspect is whether everyone answered everything as expected. It can be challenging to do this kind of pattern analysis if your survey has included branched questions, so re-arrange your data to enable to you see the patterns that are formed in the responses. Once you've got an overview of the data, it's time to drill-down and inspect individual responses. There are no short-cuts for this kind of analysis. You need to look at each textual answer one-by-one, and ask yourself whether the question has been answered, and whether the answer is adequate. For a questionnaire, responses came and can be reproduced in neat boxes. You might want to label each box in some way, perhaps use the question number and a unique identifier like a letter for the respondent. This will make it easier for you to note and find information as you progress your analysis. If you have used interviews, then it is often helpful that you tabulate them (like the Watney interview in Table 9.6 so that you can record markers—perhaps timings (duration into the interview) or line numbers. Similarly, if you have used video, you might want to note key events in the recording, and note the time into the video that these occur (Fig. 9.7), so they are easy to find again.

Analyzing the Discourse Data

Once recorded and indexed, discourse can be analyzed. Analysis of discourse can be thought of as a four-step process:

1. Read
2. Question
3. Reason
4. Understand.

Fig. 9.7 Tagging key events with times helps put together data from multiple sources (Feist, B., *A real-time journey through the first landing on the Moon*, https://apolloinrealtime.org/11/ (2019))

Table 9.7 Example of a simple table of occurrences of some themes in a survey

Theme code	Theme title	Number of mentions	Respondents including mention of theme
AV	Air vents	60	A, B, C, D, G, H, M, N, R, S, T
DB	Dashboard	32	D, E, F, G, J, K, L, M, N, Q, R, S, T
I	Instruments	24	A, B, C, D, E, F, H, I, J, K, P, Q, R, T
SP	Seating position	18	B, E, I, K, L, O, P

Once you have read through everything, accounting for things like missing answers, unrecognizable words, minor anomalies, you can then re-read questioning the meaning of the responses (or actions). This is where you start looking for repeating themes. One way to do this is to make print-outs of your data, and use highlighter pens or different coloured crayons to identify repeating themes. For example, if responses to a survey of car drivers keep referring to air vents, make a clear note of all incidents where the reference to "air vents", "air" and "vent" occurs. It will help your analysis to label these incidents, for example "AV1, AV2… AVn"; this is known as coding data themes. It might be the repeated word or phrase indicates a strength of feeling about an issue. Note that several themes may well emerge from the transcripts, and it might be that there are overlaps—i.e. one response or phrase might be categorized in two or more themes.

The simplest thematic analysis is the count of number of occurrences, which is easily presented in a table, like the example in Table 9.7. Note how the respondents are anonymized through the use of letter labels. Also, the themes have been ordered in rank of most frequently mentioned to least frequently.

Considering this table, you might conclude that "air vents" are the most important theme, given that they have the highest number of mentions. It is important to realize that the results of your investigation are not yet complete. [A student who is not going to achieve success in their dissertation might at this point report that air ventilation is the most important concern amongst all drivers. Even justifying this by abductive logic, and stating that "because air vents get the highest number of mentions, this is probably the most important issue to all drivers" is a flawed statement. From the codes of respondents listed, we can clearly see that not all drivers are concerned with air vents.] Be careful not to think that this is the end of analysis of qualitative data, feed the words and numbers into a word-cloud generator and present something that has very little useful meaning.

In the subsequent re-reading of your interviews, you need to consider causation and reasons—whether there is a relationship between themes. Also, you need to consider whether there is a hierarchy in themes. Question whether Theme #1 gives rise to Theme #2. For example, in a car-driver survey, as well as an AV theme, there are repeated references to instruments (variously described as gauges, dials, readouts or displays) that can be summarized as an IN theme. This might be different (subtly) to the theme DB, which is more about layout, size and surface finish. A way of presenting this is in a form of network diagram like Fig. 9.8. This is a rough

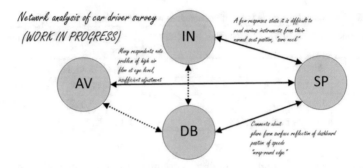

Fig. 9.8 Example of an unfinished sketch of a thematic network diagram

work, and would need some tidying up if you present it within your report. Tidying-up should include things like codifying lines—present a key to explain what the various format of solid and dotted lines mean (ranked order of "mentions", strength of relationships between themes); you will need to design your own convention.

Not all themes might occur in one network. Some themes may not connect with other themes, or there may be several discrete networks. If the design of your survey instrument has been effective, the thematic networks will align with your objectives [it is always acceptable to return to your objectives and re-write them before you submit your report for examination].

The process of plotting out a full network of themes will help develop your reasoning of the relationship between themes, and consequently your understanding of the results.

Analyzing Video
The process of analyzing video is much the same as that for interviews with the added layer of visuals added to the discourse. The technique for analysis is similar to text, involving viewing and reviewing to extract meaning and understanding. The general advice is make clear notes about what you see (or hear) that you can use to carry out your thematic analysis.

9.2.6   Discussion of Results

The discussion of results is your opportunity to explain your understanding of the outcomes of your survey. Include each theme, and explain its relevance to the objective(s) you set at the start of your project. Be clear in your reasoning, and explain whether you can deduce conclusions from what you are presenting, or whether you are relying on inductive or abductive logic to get to your conclusion.

Consider whether results can be expressed graphically. Quantitative data is easily represented in charts and graphs, but sometimes it is also possible to use diagrams to explain the relationship between various factors, or groups of results. It is also

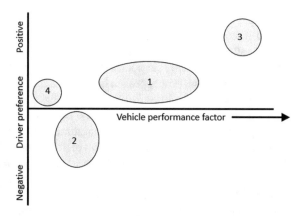

Fig. 9.9 Example of representation of opinion groups. Key: Vehicle performance factor was calculated from measureable factors (acceleration, fuel economy, etc.) Driver preference was the result of
combining responses to a survey Groups 1–4 represent the four driver types identified in the survey,
where 1 self-identified as regular long-distance commuters, 2 as regular short-journey commuters,
3 as retired former commuters, and 4 as occasional short-journey drivers

possible to mix quantitative and qualitative data in graphical format. For example,
imagine now a survey of car drivers for their preference for handling of a selection of
vehicles (qualitative) could be plotted against a compound, quantitative variable—
we'll call "vehicle performance factor"—that's been calculated using a 'figure of
merit' calculation. One of the questions in this survey asks the respondents to describe
their usual use of vehicles, and you offer them the opportunity to self-categorize.
When you plot all their preference responses on a chart of the performance factors
of the vehicles tested you note distinct clustering of results. These might best be
represented in a "blob" diagram (Fig. 9.9). The blobs represent a trend towards an
area.

The blob diagram can then support your *abducted conclusions* around each group
of drivers, for example, the group of drivers with most experience seem to have the
most appreciation for the vehicles with higher performance factors.

9.2.7 Numerical Expression of Opinion

Whilst engineering is essentially a numerical discipline, it is often necessary to obtain
and use *opinion*. This might be the opinions of an operator (computer gamer, machine
tool operator, pilot, driver) about the operation of a piece of equipment, or it might
be a general survey about user, expert or general public opinions about something
that influences a piece of design. It is common to try and find ways in which these
data can be expressed numerically—this is natural since, as engineers we're very
used to using and manipulating numerical data, but also extremely fraught and open
to misuse.

Consider for example this apparently simple question "how comfortable is this chair to sit in?", and then a response is requested on a scale from 1 "extremely comfortable" to "extremely uncomfortable". However just about anywhere but 1 and 9,the obvious extremes, will inevitably be poorly defined. So, in order to generate data which is really meaningful, the best approach is to use a decision tree—perhaps like the Cooper-Harper Handling Qualities Rating Scale at Fig. 7.3. That particular example is used for providing a numerical score for the compensation required by a pilot or astronaut opinion handling a craft, but variations can be generated for providing opinion about the steering of a car, the comfort of a chair, operator workload (a standard for this exists in NASA-TLX), the ease with which a specific task can be carried out using a computer user interface and many other problems.

If you do use methods like this to gather human opinions, and this is a very legitimate thing to do—do take care how you subsequently display and analyse the data. Some sub-fields may consider taking, for example, mean and standard deviation opinions and using those to be legitimate. Others may consider this to be inappropriate and, whilst they would be content with for example graphical display of opinion distribution, would regard any statistical manipulation to be completely inappropriate. If unsure—consult with your supervisor, whose experience in your sub-field should for the purposes of the dissertation report be definitive. A further "best practice" hint here is also, if garnering opinions, toinclude in your data some appropriate measure to define the status of the respondent: for example, if a driver, their age and number of years driving, if a computer operator, their number of hours (or years) experience in the job, if a disabled person, the nature of their disability, and so-on.

Chapter 10
Presenting Engineering Analysis

Abstract Analysis of problems using engineering theory and mathematical equations is an important engineering skillset, and needs to be demonstrated in the dissertation report. This must be done in a manner that shows the student's own abilities, and avoid reproduction of large pieces of work from elsewhere. Best practices are described and a worked example given of how to present a piece of engineering analysis.

© Springer Nature Switzerland AG 2020

P. Gratton and G. Gratton, *Achieving Success with the Engineering Dissertation*,
https://doi.org/10.1007/978-3-030-33192-4_10

10.1   Introduction

We are engineers, and we use equations as a mechanism for explaining how things work and for solving problems. By this stage in your education you have seen many hundreds of equation-based solutions and explanations, in teaching, textbooks and in journal papers. You'll see surprisingly few of these in the future as a professional engineer, but enough that familiarity with them remains essential.

Almost certainly, your dissertation will be marked against a rubric that includes a requirement for the use of engineering analysis to solve problems. This does not mean that you need to produce the sort of in-depth creation of complex equations that you'll have seen as you learn engineering science, or may have seen in some journal papers—even many PhDs don't do that. But you do need to demonstrate that you can use engineering analysis to solve at least one problem—ideally one that genuinely needed solving. This can be as simple as ratio, efficiency and mechanical advantage ratios for a machine (Fig. 10.1), or electrical loads, efficiencies, waste heat and radiation requirements.

10.2   Carrying Out, Then Writing Up Your Analysis

First, make sure that you are familiar with proof and methods that are normal in your engineering field. This is easy enough—the style of working should be consistent between your lecture notes, textbooks and the related journal papers that you've been reading through in the journey towards your dissertation. Conventions will include the degree to which you start at first principles or (far more likely) provide a reference and then work from well-established standard results.

Whilst there are many computer applications that can be used to analyse a problem (if you're not familiar with the choice look at the Wikipedia page *List of Computer Algebra Systems)*, the majority of engineers still are likely to find their preferred approach to working out how to analyse an engineering problem to be best done longhand, supplemented by simple calculating tools such as a calculator or spreadsheet. In general, have lots of paper, and a good large table to work on, and number everything you're doing so that it's easy to track back and forth through your reasoning. You'll make many mistakes, and go down many blind alleys as you work out how to solve any problem: that's fine and normal, but you need to be able to work out where that happened, and then try other ways. Don't throw away your mistakes—you may decide that they contain clues as to how you want to solve your problem later. Either keep all the paper in a folder, or save each version of your application-based attempts to solve the problem with a naming system that makes sense to you, accompanied by notes as to what you'd done and what you decide was wrong with it. This is where a project log book is useful.

You should eventually decide that you have a workable method to solve the problem in hand, which uses the theory you have been taught and/or found and developed

Fig. 10.1 Machine requiring careful calculation of loads, gear and thus speed and torque ratios, and efficiencies: hence input work (Illustration of a hypothetical specialist snow plough by Mr William Heath Robinson)

in the course of your dissertation project. With either longhand or computer analysis, you would have worked that through and found the solution that does what you wanted. You then need to write it up for inclusion in your interim or final reports.

10.3 Writing Up Your Engineering Analysis

When you write up your analysis ensure you write it with a good narrative flow allowing your reader to clearly follow the train of argument from problem to solution. Ensure that a competent engineer in your discipline could follow it through. If, to ensure you've displayed your ability to perform competent engineering analysis you've had to contrive a problem somewhat—don't worry, everybody knows that's part of the dissertation "game". Equally however, if you have actually done multiple pieces of similar analysis, don't be afraid to use terms like "following a similar method to that in section NN, the following was determined", as repetition of description becomes unhelpful and can create accusations that you are padding your work out unnecessarily.

Number everything consecutively—prefixing with chapter or section numbers is normally a good practice rather than rolling over numbering from chapter to chapter. Typically, you'll have separate numbering values for equations, figures and tables. Ensure that in the description of your method you always refer to the figures, equations and tables by number so that your intended narrative flow is clear. Always ensure that you've stated your assumptions, and provide references to information you've drawn upon.

There is often a temptation amongst dissertation students to re-type or re-use large pieces of prior proofs or analysis from, for example, papers or textbooks. Virtually never is this wise or acceptable—at best, it will annoy the academics marking your work; at worst, it can create accusations of plagiarism. Best practice is simply to identify either the earliest, or the most commonly used form of that original method, and cite it. If you've changed any of the symbols or methods from those in the original, do state that, and what your changes are—but at that point stop and just present your own work.

Check whether your institution's writing guidelines require you to include a list of figures, equations and so on. If they do, make sure you've done so in the approved manner (as you should also have done for symbols and abbreviations, which will certainly be required).

10.4   An Example of How to Write Analysis—Some Take-Off Distance Calculations

This is taken and simplified from one of the authors' own research papers.[1] Whilst technically from a specialist application, it has been used here as it is a reasonably simple engineering analysis that doesn't require any specialist knowledge to follow. Note the way it has been written, the use of references, the explanation of variables, and use of a simple diagram.

> It is conventional to divide the take-off into three distinct segments: the initial ground roll, the post-rotation ground roll and the climb to screen height.
>
> Complying with normal certification practice, which requires at least six data points [1], six take-offs were carried out, the results (times, speeds) were tabulated, and then distances are calculated as shown below.
>
> The method makes the following assumptions:
>
> - During each segment, the aircraft's acceleration/deceleration is constant.
> - Surface wind velocity and direction are constant between ground and screen height.
> - During the air segment, the aircraft climbs in a straight line between the unstick point and screen height: for microlight aircraft and smaller light aeroplanes particularly this is a reasonable assumption, since the initial climb condition for light and microlight aeroplanes is normally established within 1.5–3 m of the ground, which is small within the 15 m climb to screen height. Where an aircraft's curved transition from rotation to steady climbing flight comprises a larger part of the climb to screen height (particularly likely to be the case for an airliner where screen height is normally 10 m, pitch rates are lower, and inertia greater), then it is likely that this assumption will require re-consideration, probably by comparison with aircraft measured flightpath using GPS, kinetheodolite or video analysis methods. To date, however, this has not been found necessary.

The following notation is used (see Fig. 10.2): speeds at start, rotate (pilot pulls back on stick causing nosewheel to leave ground), unstick (aircraft takes off) and screen (minimum safety) height are 0, V_1, V_2 and V_3, respectively. Surface wind is V_W and is positive when a headwind is encountered. Times of each segment are t_1, t_2, t_3, respectively. The lengths of each segment, measured along the ground, are S_1, S_2 and S_3. Straight line distance from unstick to top of screen $= S'_3$. Accelerations during each segment are a_1, a_2, a_3 (a_3 is acceleration along flightpath, not along the ground). Screen height is h. For calculation, all the above will be in SI units (m, ms^{-1}, s).

To determine the length of the first (pre-rotation) ground roll segment:

Assuming that the aircraft is initially stationary,

$$S_1 = \frac{1}{2}a_1 t_1^2 \qquad (10.1)$$

[1]Gratton, G.B., "A timed method for the estimation of aeroplane take-off and landing distances", *RAeS Aeronautical Journal* Vol 112 No 1136 (Oct 2008), pp. 613–619 https://doi.org/10.1017/S000192400000258X.

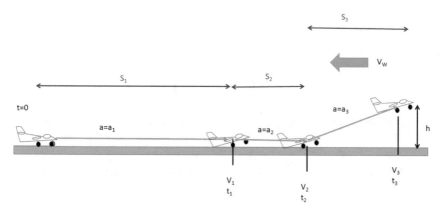

Fig. 10.2 Illustration of take-off segments

$$V_1 - V_W = a_1 t_1 \therefore a_1 = \frac{V_1 - V_W}{t_1} \qquad (10.2)$$

(Noting that the aircraft translational velocity used in calculation is $V_1 - V_W$.)
Inserting (Eq. 10.2) into (Eq. 10.1) gives:

$$S_1 = \frac{t_1}{2}(V_1 - V_W) \qquad (10.3)$$

To determine the length of the second (post rotation) ground roll segment, we use:

$$S_2 = (V_1 - V_W).t_2 + \frac{1}{2}a_2 t_2^2 \qquad (10.4)$$

$$V_2 - V_W = (V_1 - V_W) + a_2 t_2$$
$$\therefore a_2 = \frac{V_2 - V_1}{t_2} \qquad (10.5)$$

Inserting (Eq. 10.5) into (Eq. 10.4) gives:

$$S_2 = (V_1 - V_W).t_2 + \frac{1}{2}\left(\frac{V_2 - V_1}{t_2}\right)t_2^2$$
$$= (V_1 - V_W).t_2 + \left(\frac{V_2 - V_1}{2}\right)t_2$$
$$= t_2\left(\frac{V_2 + V_1}{2} - V_W\right) \qquad (10.6)$$

To determine actual length of the air segment, we obtain:

$$S_3' = t_3\left(\frac{V_3 + V_2}{2} - V_W\right) \qquad (10.7)$$

But by Pythagoras:

$$S_3' = \sqrt{S_3'^2 - h^2} \tag{10.8}$$

and inserting (Eq. 10.7) into (Eq. 10.8)

$$S_3 = \sqrt{t_3^2 \left(\frac{V_3 + V_2}{2} - V_W \right)^2 - h^2} \tag{10.9}$$

total take-off distance to screen height, in actual conditions, is then $S_1 + S_2 + S_3$, as determined above, so:

$$S = \frac{t_1}{2}(V_1 - V_W) + t_2 \left(\frac{V_2 + V_1}{2} - V_W \right) + \sqrt{t_3^2 \left(\frac{V_3 + V_2}{2} - V_W \right)^2 - h^2} \tag{10.10}$$

Or in slightly modified form:-

$$S = \frac{t_1}{2}(V_1 - V_W) + \frac{t_2}{2}(V_2 + V_1 - 2V_W) + \frac{1}{4}\sqrt{t_3^2(V_3 + V_2 - 2V_W)^2 - 2h^2} \tag{10.10a}$$

Adjustments to standard conditions may be made using the usual variance factors [2]. However, this method does not take account of the errors which normally exist in the variables. It is assumed that each of these factors is accurate within the precision of recording (which is normally done manually, by a Flight Test Observer (FTO) or the Test Pilot themselves).

References

[1]   UK Civil Aviation Authority, *British Civil Airworthiness Requirements, Section S, Small Light Aeroplanes*, CAP 482 issue 3.
[2]   UK Civil Aviation Authority, *Take-off, climb and landing performance of light aeroplanes,* AIC 127/2006 Pink 110.

Chapter 11
Drawing and Presenting Conclusions

Abstract All engineering reports require conclusions, which must flow from the narrative in the main body of the report. The nature of the dissertation will dictate the form of the conclusions, and here are given explanations for the conclusions from the main forms of dissertation: a computational project, a design-built-and-evaluate project, a feasibility study project, an experimental research project, a theoretical research project, and a test and evaluate project.

© Springer Nature Switzerland AG 2020
P. Gratton and G. Gratton, *Achieving Success with the Engineering Dissertation*,
https://doi.org/10.1007/978-3-030-33192-4_11

11.1 Introduction: About Conclusions

"Conclusions" is one of the shorter chapters of your dissertation report, but also one of the most important. It can have a major impact, both on the value of the project and on your marks at the end. If the conclusions are missing, some markers will be well within their rights to give you a fail mark. So the conclusions need far more thought, work and effort than their short few pages would suggest.

The whole dissertation report has led to the conclusions, which in themselves may contain new original thought about the work, but must never contain any new information. So, any references within the conclusions should be entirely internal back to earlier parts of the report, and not to external references (except where essential to the explanation and have already been used, so that a previously made point is just being reinforced).

Whilst of course you should put a lot of effort into the quality of your written language throughout, this is also one of two sections: the other being the abstract, where you should put as much effort as you can afford into getting the spelling, punctuation and grammar as good as you are able to make them.

11.2 Type of Conclusions

There are, of course, different types of dissertation projects and reports, so it would be surprising if all conclusions needed to read the same. We've tried here to show how you should construct your conclusions for each of the main types—clearly, if you're unsure, this is one of those things you should be talking to your supervisor about, and asking whether they can review your structure and content. Whilst we've shown six main forms of dissertation, your own project may diverge from—or more likely combined aspects of these (e.g. a project which carried out modelling or analysis, then used experimental work to validate it)—things that don't necessarily fit into nice neat categories. Your university should have guidelines as to the preferred length (most guidelines are likely to suggest 2–3 pages for the conclusions: so if there aren't any guidelines at your institution, then choose a reasonable length), and possibly structure, of conclusions—make sure that you're familiar with them and stick to them rigidly: tolerance for overlength or (in the opinion of the markers) irrelevant conclusions is likely to be poor. Also think forward—in the coming months you will almost certainly be making an assessed presentation about your project, and after that hopefully be sitting in interviews for a job, or for a higher degree: this concise conclusion may well be something you're asked to talk about, and both critique and defend: so the work you're doing now is preparation for all that as well.

Concluding a computational project dissertation report
The first part of the conclusions of a computational project dissertation is likely to summarize what you were able to do, along the lines of:-

This project created an application in (language/coding tool) to run in (hardware) that was able to (problem to be solved). It did this by… . The resultant application was (summary of structure) and able to (summary of actual functionality).

The next part of the conclusions is likely to explain the outputs of the project: what it was possible to achieve. This should be factual, showing both the strengths and weaknesses of the product: almost certainly there were limitations to what the computational product was able to achieve, or limitations to the quality of the outputs and these need explaining. Keep all of this factual, and make quite sure that everything has been supported by data and analysis in the main body of the report. At the same time, there is absolutely no requirement to "beat yourself up": informed criticism is about what happened, why and what might be improved, and is not about blaming yourself or any colleagues you were working with for deficiencies in product, or even if it's failing to work at all.

Concluding a design (build, evaluate) project dissertation
Designing, building and evaluating something, whether it's a mechanical device, an electronic system, a building, a model or a piece of software, is absolutely at the root of what an engineer should have been trained to do. However, this is also amongst the hardest to conclude, and the conclusions section of such a project report is likely to be in large part comprising a summary of (a) the design specification, (b) the design (and if applicable, manufacturing) methods used to meet that specification, (c) a brief description of the solution, (d) how the quality of the design solution has been assessed, and very importantly (e) how this could all be improved upon in a future solution.

While writing a conclusions chapter like this, there are two main traps to avoid. First, don't let this become too lengthy: the bulk of what has to be said about your specification, design and build should have been in the main body of the report, and the conclusion is just concluding what should have been a coherent flow of description through the main report. Secondly, never bring in new information to the conclusions: there must be nothing in the conclusion which couldn't be deduced by a skilled reader from a careful reading of the main body text. The conclusion must be your own expert (and you most certainly should be an expert by now) summary of the work only, and nothing else.

Invariably, your assessment of what you've designed will include identification of significant deficiencies in what's been done. This is absolutely inevitable, and you won't help yourself by fixating on them: you were one student on your own, or part of a small team, with limited resources, performing a task that was to you largely new. Dwelling on the problems will annoy the reader, but ignoring them will annoy them more. So, keep the description of what is strong, and weak, about the product factual and to the point, don't allocate blame to people (only methods and the limitations of science, technology and resources) and keep this relatively brief.

Concluding a feasibility study dissertation
In doing a feasibility study you would have studied how an engineering solution may be achieved and how. You are likely to have also been analysing costs, timescales and risks, as that is, of course, the whole nature of a feasibility study. If you have

found this work, compared to classical engineering analysis or design work, to be very new and unexpected, that is understandable but that discussion doesn't belong to the conclusions: there should be a separate "reflective" section for that.

Writing the conclusions to a feasibility study should be relatively straightforward. The conclusions section should be ordered along these lines:-

- A very short explanation of what the feasibility study set out to establish, and the resources used to do that.
- An explanation of the extent to which you've concluded the specification can be met (or if not, why not).
- A summary of the costs and timescales predicted to be required to deliver the project.
- A summary of the technical risk associated with delivering the project. Technical risk in this context mainly amount to a set of explained "error bars" on the time, human and capital resources needed to deliver it, but also any major factors which might prevent delivery.

As with all conclusions, this should all be brief, and everything is built upon the body of the report. What you are essentially doing with the conclusions here is saving the reader, at their first reading, from going through and interpreting for themselves all the detail of the report to understand the conclusions themselves. Of course, the expert reader (including, of course, anybody marking it) will still be doing that—but the conclusion (despite coming at the end) provides them with a useful key—as well as the duality giving them a chance to second guess whether they agree with your conclusions.

Concluding an experimental research dissertation

An experimental project is likely to be the most familiar ground for most students, as it's an extension of a classical laboratory report which you've been writing since being at school. At every stage you have of course progressed to yet more complex reports, and this is just the latest and most complex form.

So, your experimental research conclusions should look very much like those of the many lab reports you've previously written—longer and more complex of course, and the imperative for it to follow from the body of description, results and analysis is as imperative as it has ever been. The main necessity here is <u>not</u> to try and be too original in how you write it. Stick to tried and tested approaches; however, be. respectful of the fact that what you've done with your dissertation is massively larger than any previous task you've undertaken. If you look at academic journal papers in your field, you'll see examples that show this trend again (although a journal paper is normally much shorter, and assumes much greater prior knowledge by the reader: unless your supervisor tells you to, mimicking that particular style may not be a good idea).

You may have done this project with a research hypothesis in mind, in which case this should have been stated in your original introduction. If so, a very important part of this conclusion is an explanation of whether the hypothesis was confirmed or disproved—and if so, with what degree of confidence.

Concluding a theoretical research dissertation

Theoretical research, like any other research, should have been written up in a systematic manner: starting with the research question, through the literature review, to the data used and analysis carried out, and then to the conclusions. The conclusions should read very similarly to those of an experimental report (above) although experimental conditions are replaced by assumptions, and results described become the results of literature review and theoretical analysis, not of experiments. As with experimental work, there may have been an opening hypothesis to the work, and if so, a very important part of the conclusion will be explanation of whether the hypothesis was confirmed or disproved—and if so, with what degree of confidence.

Concluding a test and evaluate dissertation

The conclusions of a T&E dissertation are often the easiest to write, because the whole point of the dissertation is to find the conclusions. It's likely that you'll want to use the following order to describe them

(1) This project has evaluated (very short description).
(2) For the purpose of (what is it for, or what assessment criteria were in use).
(3) At the following conditions (temperatures, pressures, speeds, operating environment, trial subjects, etc).
(4) Explain the items found unacceptable, and why.
(5) Explain the items found unsatisfactory, and why.
(6) Explain what was "enhancing" and should be considered for re-use on other artefacts.
(7) In summary, was the overall system acceptable, unsatisfactory, unacceptable (or other scoring terms as appropriate to your discipline and task)?
(8) Describe whether the system as assessed complied with the relevant engineering standards, and if not what specific non-compliances were found.

Take care to separate out standards compliance and actual assessed suitability of the item under evaluation—it is not uncommon for items to meet standards but be unfit for purpose, or alternately very good, but not in compliance with some narrow aspects of an engineering standard. The reader must be left clear as to the differences.

11.3   Further Aspects to Conclusions

Some universities' guidelines will also include in conclusions: recommendations (for further work or for other activities) and reflection on how a project went. In this book, we've assumed that these are separate—and they're covered by the next two chapters; however, as always, be guided by your universities' guidelines on how they want your report structured.

Chapter 12
Making Recommendations

Abstract It is common, and sometimes essential, that an engineering report contains recommendations, which are normally immediately after the conclusions. Recommendations may either be for further work (e.g. additional research into a topic identified in the project) or for action (e.g. to rectify problems with a piece of equipment evaluated during the project). Recommendations should normally be classified according to how important it is that they are carried out, with various terms and scales available for this purpose; two of which are explained here.

© Springer Nature Switzerland AG 2020   171
P. Gratton and G. Gratton, *Achieving Success with the Engineering Dissertation*,
https://doi.org/10.1007/978-3-030-33192-4_12

12.1 Introduction: Report Recommendations

The dissertation, like many engineering reports, should normally contain recommendations—likely to be at the end, either within or after the conclusions. Normally, they'll also be contained in short form at the start, in the abstract. To maintain good narrative flow, they must always lead on from the body text, through the conclusions. Making recommendations that are not well supported devalues the report, and discredits the author as a (present or future) professional engineer.

Recommendations should normally be brief and to the point, because the reader who wants the detail and supporting information will get those from the body of the report. The stated recommendations are very much headlines: albeit that there are certain conventions about breakdown and justification which matter.

Recommendation sections, therefore, should normally be short. Don't, however, let that brevity leave you spending only minimal time on it. This is a very important section of any report, and readers will put much store by them—take time to weigh every word of every recommendation, and ensure that it is a really useful final part of the core report.

12.2 Types of Recommendation and How They're Written

Recommendations can be categorized into two distinct types. If your report gives rise to both types, it is good practice to present these as separate sets. The types of recommendation are:

- recommendations for further work
- recommendations for action.

Recommendations for further work
Particularly in research-oriented reports, it's common to make recommendations for future research to be done. Where the work has identified significant deficiencies in knowledge that can be built upon with future research, recommendations can point the reader towards what research can be done to plug those knowledge gaps.

For example, perhaps your dissertation may be into the design of a touch-screen learning package to help 10 to12-year-old children learn about the basic principles of temperature measurement. Temperature measurement is important of course in many fields, but you discover during your literature review that there's actually no useful consensus about what prior knowledge about temperature measurement is most useful before going on to study secondary or high school science. You might recommend that there would be value (without necessarily saying who should do this, or how) in research being done to establish what those learning outcomes should include. Alternatively, consider an example where you have been trying to design a product using a new composite material, but discovered that there is no good data about its properties at elevated temperatures; so, a recommendation that there is a need to characterize elastic modulus and tensile strength over a particular range of temperatures follows naturally from your work.

Recommendations for action

Recommendations for action are somewhat different to recommendations for future work, and in particular to recommendations for further research (which is the most likely "future" work recommendation you may write). Action recommendations are written because something has specifically been identified that bears improvement, for example:

> This research has shown that the light switches in this building are too high for 70% of the residents to reach, it is recommended that they are lowered by 400–600 mm, making them reachable by all residents (Highly desirable)

In those engineering disciplines that specialize in assessment of products (e.g. flight test engineering), it is normally regarded as good practice that the engineer who makes recommendations should take care to own the problem, and not the solution. So, the example above is essentially saying that the switches are 400–600 mm too high, which is a statement of the problem. Providing a clear statement that the switches should be located to positions X, Y and Z is owning the solution; that may be fine if the report is a design report, but not if it is an assessment report. Design recommendations and product assessment are conventionally done by different and independent specialists. Be clear which role you are taking.

Some disciplines will classify recommendations, in terms of how important they are. If you have been given, or your discipline of engineering has a standard scale—or you want to create your own—that is likely to be fine: just make sure that either your main report, or a reference, contains a clear definition of the scale you are using. Otherwise the following terminology is often used and intuitive:-

- [This recommendation] is considered ESSENTIAL (usually following on from a conclusion that the deficiency being addressed is UNACCEPTABLE).
- [This recommendation] is considered HIGHLY DESIRABLE (usually following on from a strong conclusion that the deficiency being addressed is UNSATISFAC-TORY).
- [This recommendation] is considered DESIRABLE (usually following from a less strongly worded, *unsatisfactory* conclusion, usually following from an observation that a particular characteristic, whilst okay, could be much improved.)
- (Much harder to write), [This recommendation] is considered ENHANCING and implementation is recommended.

Take care not to go overboard with recommendations, in particular *essential* recommendations which are a very powerful tool, not to be overused—and usually reserved for applications where the recommendation is related to a risk of injury or death, otherwise their use tend to destroy the writer's credibility. More subtle numeric scales, perhaps a scale of 1–10, are another approach, which may be particularly useful when a report might be used by managers subject to fixed budgets, and thus needing to prioritize spend. This can be an excellent solution, but only if the scale is clearly defined and explained. The following is a 7-point example, but may only be appropriate to some kinds of product—and does not have its basis in any published standard, so you may well be better off creating your own different standard:

#1—If not rectified, may lead to loss of life
#2—If not rectified, may lead to serious injury
#3—If not rectified, may lead to minor injury
#4—If not rectified, the product may not work
#5—If not rectified, the product will operate sub-optimally
#6—Rectification will resolve minor inconveniences
#7—Implementing this recommendation will enhance the product.

Don't be afraid to have barely used extremes in such a scale—for example, the 7-point scale above would, for teaching software evaluation, seldom be used outside of the range 4–7, whilst for a safety harness might usually only be used in the range 1–5; yet, the same range could conceivably be used for both, making it particularly useful (Fig. 12.1).

Fig. 12.1 Don't forget when constructing recommendations, to fully consult with and enlist the help of product end users

Chapter 13
Reflection

Abstract It is common best practice after completion of a project to formally reflect upon how it went, performance of participant(s), and what lessons can be learned from the project's conduct. This chapter describes this particularly in the context of a dissertation, with a suggested approach to both considering and formally recording these reflections, which may be required by many universities.

© Springer Nature Switzerland AG 2020
P. Gratton and G. Gratton, *Achieving Success with the Engineering Dissertation*,
https://doi.org/10.1007/978-3-030-33192-4_13

13.1   What Have You Learned from Doing This Project?

It is common for university project reports to require a reflective piece to demonstrate the personal investment that has been made in time and effort in getting it to the point of submission. This reflects what happens in industry, where there is usually a "wash-up" meeting or survey after the completion of a project, where all parties involved in the project can reflect on their learning points; this is sometimes followed up with a "post-completion report". Every university, or every engineering course, will have its own preference for how this is done—for some it may be a short section or a separate chapter within the report, whilst others might require a separate report, or have a systematic learning log that students use with every piece of coursework they submit.

If your university doesn't have an established system, we propose that best practice is to have a chapter within the report (so all your report's readers can clearly identify it), perhaps towards the end, after the references, but before the appendices—the position is somewhat irrelevant, the point is that this is something you can only write after you have completed all the work in the project, including writing the report.

13.2   Lifelong Learning, Part of Being a Professional

The engineering profession has always encouraged the notion of lifelong learning, also known as *Continuous Professional Development* (CPD), as best practice. In order for CPD to be effective, personal reflection is required, both in action (whilst you are carrying out something), and on action (when you complete something and have the chance to look back).[1,2] Medical students are encouraged to undertake reflection from the outset of their medical education. This is because they have to learn to deal with life-and-death situations that, in all probability, will occur in their professional lives; reflection helps to develop emotional resilience. Because engineers don't generally tend to deal with life-and-death on a day-to-day basis, reflection has not always been "taught" as a learn-able mode of operation, and individuals are left to work out for themselves how best to go about it. This isn't good enough, so let's consider a framework for reflection.

---

[1] Solent University, "How to think reflectively: Schön's model", *Archived Material (Reflection)*, (2012–2019) Available at https://learn.solent.ac.uk/mod/book/view.php?id=2732&chapterid=1113 [Accessed Aug. 10 2019].

[2] Schön, D. A., *The Reflective Practitioner: How professionals think in action*. London: Temple Smith (1983).

Fig. 13.1 Diagram of
Gibbs' reflective cycle

13.3 A Model for Reflection

The easiest way to "do" reflection, and to write your reflection, is to use an established framework. There are many of these, and there are also many interpretations of them. What we suggest here is known as Gibbs' Reflective Cycle.[3] It is a simple six-step approach, so it's easy to learn and apply in all sorts of places, not just about reflecting on your dissertation project.

The six stages of Gibbs' model are:

Description
Feelings
Evaluation
Analysis
Conclusion
Action Plan
… and these are usually expressed as a cycle (Fig. 13.1).

Description
Obviously, this is what you did. In the context of your dissertation project, this part should describe briefly what you did in contrast to what you set out to do at the start of the project plan. *Briefly* is an important word, as the details of what you did are clearly reported in the rest of the report—there is no point regurgitating what is clearly described elsewhere.

This segment of the reflective piece must be factual, dispassionate and without judgment.

---

[3]Gibbs, G. *Learning by doing: a guide to teaching and learning methods*. Further Education Unit. Oxford Polytechnic: Oxford (1988).

Many undergraduates become distressed that their project does not work out the way it was intended. Perhaps their experiment doesn't work, or their model does not converge, or they don't get the results they'd hoped for. There is nothing to be ashamed of if this happens. What is far more important than results is that you can demonstrate that you understand the underlying science and processes behind your project. So, it might be that the aim and objectives you started out with do not endure for the final report!

Questions to ask yourself:

– What did I do?
– What are the facts of what happened in doing my project?

Feelings

This segment of the report is about your reaction to what happened in your project; what did you feel about how things went and how they turned out. Don't try to analyse your reaction yet (this will come later); just putting emotional reactions at various points in your project into words can be a challenge.

For engineers, this is often one of the most difficult aspects of reflective writing. There is a tendency to think that everything you write as an engineering report has to be cold and dispassionate, as though you are like the part-Vulcan, Mr Spock in the 1960s TV series version of Star Trek. It's to be remembered that Mr Spock was the Scientific Officer, whilst Mr Scott was the Engineer on the USS Enterprise: write more like Scotty. Would the 2018 portrayal of Shuri in the film Black Panther be a better role model for authoring this section? Perhaps so.

It is acceptable in this segment of the reflective report to record enthusiasm for the project topic, or disappointment for the results obtained. It may be difficult to find the right words to adequately express how you feel about aspects of the project; however, all the time it takes is an investment in learning about yourself and your relationship with the project.

Although you are writing here about emotions, this is NOT an excuse to use colloquialisms, which can be confusing to someone from another generation or different cultural background. Nor should you use emojis in this or any other part of your report. These are ambiguous and will undermine your professional credibility as an engineer.

Questions to ask yourself:

– How did I feel about tackling this project?
– What was my emotional reaction to what happened during the project?

Evaluation

This is the segment of the reflective report where you express value judgments. State clearly what you consider was good and bad in carrying out the project.

Questions to ask yourself:

– What went well in the project; e.g. followed the project plan exactly?
– What diversions from the project plan did I have to make?
– What was the easiest aspect of carrying out the project?

- What was the most challenging thing about the project?
- To what extent did the outcome of the project meet expectations?

Analysis

This is where you try to make sense of the experience of doing the project; why you reacted as you did; and why you made judgments as you did.

This segment is an opportunity to think about how factors outside the project impinged on its progress. Whilst this is not about making excuses, or apportioning blame, it can be the occasion for analysing reasons that contributed to the way things turned out. For postgraduate students, this is also an opportunity to compare and contrast your experience in this project with that of your undergraduate course.

Questions to ask yourself:

- If I deviated from the original aim, objectives or plan, then why so?
- Thinking about my feelings at various times throughout the project, why did I feel that way?
- Considering what I judged to be good about the project, why was this so?
- With regard to what I found challenging during the project, explain why?

Conclusion

The conclusion in this reflective report should be specifically about you with respect to this project; what the conclusion of your analysis is with regard for how you carried out this project. It might be about what you have learned in terms of content, or in terms of the way you work. There may be some more general conclusions that you want to bring in. This is acceptable, especially if they are lessons that others can learn from (e.g. the process required for a procurement system).

Questions to ask yourself:

- If I was starting out on a project, what would I do differently to what I did on this one?

Action Plan

This should be a personal action plan. You might think of it in two ways: what might you do differently, if you were to do a similar project in future (e.g. as a part of a Masters degree or PhD, or in a work environment), and what lessons in ways of working might help you in your future engineering career.

If you already have plans for your immediate future, it is suggested that you filter them in here to make the reflection more authentic.

Questions to ask yourself:

- What are my immediate plans?
- How might I use what I've learned on this project in my future career?

Using Gibbs' cycle of reflection as the framework for your reflective chapter, it might be possible to compose something coherent and insightful in as few as six or seven sentences.

13.4   The Role of Reflection in Group Projects

Working together with other members of a team has its own challenges. If you are lucky enough to have been part of a well-organized, cohesive team, then reflecting on the experience of working together may well be an enjoyable experience. Note that the reflection still has to consider why it worked so well, and have the balance of view, perhaps acknowledging where it might have had problems, but avoided the problems escalating.

If working in the team was not a wholly enjoyable experience, the reflection upon the learning experience should NOT be turned into a complaining or moaning piece. It does need to be honest, but needs to be respectful of the other members of the team.

When you write about the contribution of team members, you should adopt a consistent system of naming them. For example, you might just use their initials or roles (your supervisor is going to know who is in your team), or consistently abbreviated names. Do not mix styles, so that for example some people are referred to by surnames and others by nicknames—this can be perceived by your reader as being indicative of a bias towards or against particular members of your team. It is important that you make reference to each and every member of your team, even if it is to report that they stopped coming to meetings after the second week. It is important to acknowledge, for example, whether the withdrawal of a team member has impacted organization and operation within the team.

One of the frequently asked questions about writing up reflection of team projects is about how to deal with diversity in the team. Many characteristics will vary amongst team members, but most often their variety will not affect how the team works together. We recall reading some years ago a student evaluation of the Formula Student Team group dynamic—overall, 30 students were involved, three of whom were women. The male team-member who had written this report noted that he had been "worried at the start of the project that the girls wouldn't join in, but... they had all been valued members of the team and never seemed clique-y." Whilst this was a valid insight, use of the word "girls" was unnecessarily infantilizing, and we doubt that he would ever have referred to the men in the team as "boys".

Differences manifest in different ways, and sometimes the problems that arise are not always obvious. A colleague described the breakdown of group dynamics in a team of six male students to whom he was supervisor. Two were locals, who had grown up attending different day-schools, a third was from somewhere on continental Europe, but again had lived at home and been educated at a day-school, the other three came from three different countries outside Europe, but had each previously attended different boarding schools. It was observed that the team operated in two factions—the "boarders" and the "day-schoolers"—that barely communicated. Each team members' reflection seems to report only on their own sub-team, without acknowledging the other half's contribution.

13.5   How is Reflection, or a Learning Log, Assessed

As said earlier, each engineering course is likely to have its own preferred way of dealing with the reflection on learning in the dissertation project. Some courses might want to see a student's log book—and this might actually be an online log, rather than a physical book. Often a university will provide training in critical thinking and reflective writing, and might set out what they expect the reflective report to include, and how they want it written. Usually, reflective writing is presented using first-person singular pronouns, and includes the past tense (goals set; what happened), present tense (conclusions drawn; lessons being learned) and future tense (how lessons influence future aspirations).

It is quite usual to require that in addition to the final report about the project that a Project Management Report is submitted. This might include a commentary on the schedule of the project programme, and any deviation from it.

Whatever format the reflection takes, the content of the report or log is likely to be assessed on evidence of personal development. Some of the questions that are in the mind of the assessor are likely to include:

– Is this authentic?
– Is the student clear about what they have learned?
– Is the student using examples of how they might use these lessons in future?

Chapter 14
The Output

Abstract The most identifiable output of the dissertation is the report, which is likely to be the largest piece of written work yet produced by the student. This section describes the process of creating that report, how it should be structured and the emphasis on the various parts. A further possible output is a presentation, which many universities require students to make late in their dissertation and often attracts a formal mark: guidance is given on good practices for the presentation. A few dissertation projects may also create forms of intellectual property (IP) which students or the team around them may wish to exploit or protect; IP is described in the particular context of the dissertation.

© Springer Nature Switzerland AG 2020
P. Gratton and G. Gratton, *Achieving Success with the Engineering Dissertation*,
https://doi.org/10.1007/978-3-030-33192-4_14

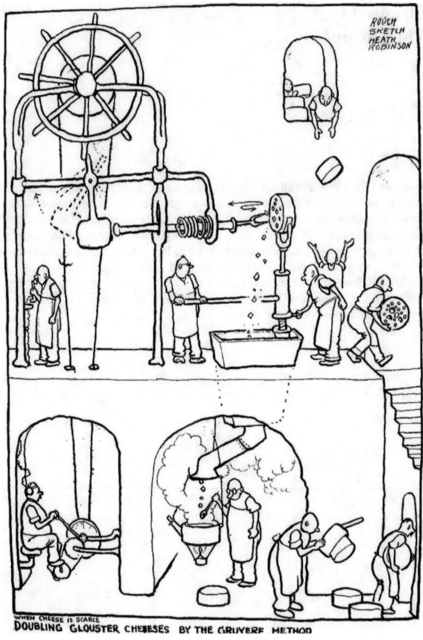

ROUGH
SKETCH
HEATH
ROBINSON

WHEN CHEESE IS SCARCE
DOUBLING GLOUSTER CHEESES BY THE GRUYERE METHOD

14.1   The Final Report

14.1.1   Overview of the Structure of the Report

The structure of a dissertation project report is always going to be similar—an introduction, a mid-section where the details of activities are explained and a concluding part that discusses what was found and what it means. Beginning, middle and end.

Obviously, it is not that straightforward, otherwise we'd not write a book about it. There are various things to be included in each part of the report, and these will be discussed in this chapter. However, the general rule is that the engineering dissertation project has at its centre an element of unique work, and so, as with any experiment, the final report will contain enough information to permit the verification of what has been done.

The general "shape" of the final report is likely to be "fatter in the middle" where the method of whatever the project features is carried out, and analysed, and is "thinner" at either end where precepts are set out and concluded. This can be roughly equated to numbers of pages allocated to the parts of the report (Fig. 14.1).

14.1.2   When Should You Start Writing the Final Report?

Before getting into the thinking about the detail of the report, perhaps we should address WHEN you should write the report. Logically, you might think that you should complete the project, and then write it up; however, it is a big project and there is a lot to write-up, so you might start the write-up of the introduction as you move into carrying out whatever the middle section is about.

14.1.3   How Should You Write the Report?

A frequently asked question is "what tense of speech should the dissertation report be written in?" As it is a report of a completed piece of work, it should be written in a past-tense.

14.1.4   Contents of the Final Report

The typical report contains the following:

– Title page
– Abstract

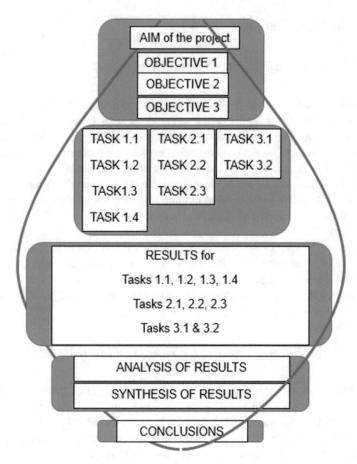

Fig. 14.1 A typical project report "shape"

- Contents
- List of figures
- List of tables
- List of notations, with units
- Introduction
- Methodology
- Results
- Discussion
- Conclusions
- Suggestions for Further Work
- References
- Appendices

It is recommended that each of these headings is treated as a separate chapter, started on a new page. The following sections explain what each chapter is about, and what it needs to contain.

Title page
Obviously, this contains the title of the project, but it should also usually contain the title of the degree course, and the university. It should probably contain your name (unless your university has a rule about anonymous submission of work) and unique personal identifier (student number). It is also helpful to have the date of submission, or at least which academic year. There may be a standard template you should use—check whether this is the case.

Abstract
This should be made up of one paragraph with usually three parts. The first should explain what the project is about, and why it is significant. The second part states the methodology used to carry out the project, including how any data is collected and analysed, with particulars about theory and software applied. The final part is to provide the highlights of the results and the conclusion of the dissertation.

Contents
Number your chapters, and sections within your chapters (and sub-sections within those sections; for sake of clarity, we suggest that you don't have any more than four levels in your contents hierarchy, and three often read better). In the Table of Contents list, you might only show chapters and sections with their page numbers to help your readers navigate their way around. The usual convention is for Chapter 1 to start on page 1. All the pages before Chapter 1 are commonly numbered using lower case Roman numerals (i, ii, iii and so on), as is the case with this book.

List of figures
Each figure within your report must have a unique identifier and a title. You may just start with Figure 1 for the first figure and continue in number order throughout the report. Alternatively, you could use the chapter number as a prefix; so, the first figure in Chapter 3 is Figure 3.1. Whatever system you adopt, ensure that you apply it consistently throughout your report. The list of figures should normally start on a new page.

List of tables
Whatever system of numbering figures you adopt, it is usual to use the same system to identify tables. So, Table 3.5 is the fifth table in the third chapter, and so-on (again, like this book). The list of tables can usually be on the same page as the list of figures, so long as both are short enough.

List of notations, with units
This should include any symbols you use, and any acronyms. This should include everything, even the very common ones we use in everyday speech. You need (picking an example of non-engineering terminology that may come up) to decide what you are going to use for things like carbon dioxide. CO2 is commonly used, but to be accurate, this form of notation is actually incorrect. The chemical symbol is CO_2

using a subscript for the number two. Whatever you decide to use, make sure it is used consistently throughout your report—including in diagrams. This list is likely to be extensive. It is recommended that you start it on a new page. You might consider listing acronyms separately. This decision depends how many you include in the list and how easy you want to make it for your reader to look up.

Introduction

This is Chapter 1 of your report, and should normally start on page 1. It is highly likely that this chapter will have a number of sections—starting with an explanation of why this project is important and what merits it being carried out. There might be some context that will be helpful to your reader in understanding the project.

A key issue to address here is what has been already published about the topic of the project. This is usually referred to as a "literature review". It is not necessary to have a chapter (or a section) entitled Literature Review. It is totally unacceptable, however, to have a report about a project with absolutely no evidence of reading around the topic. It is important to acknowledge, as far as possible, what work has been carried out in this area and how your project fits in with this. A frequently asked question is "how many publications should a good literature review contain?" Every project is going to have its own answer. You need to provide evidence of a breadth and width of reading around your topic. This means including support from a broad cross-section of types of publications, including, ideally, some published research from peer-reviewed journals, perhaps some guidance documents, technical notes and conference proceedings from learned societies; then, depending on the nature of your project, case studies and opinion pieces from the engineering trade press (refer to Chap. 5).

In various types of engineering projects, there will be undoubtedly some scientific theory. You might cite textbooks for this kind of background. In the introduction chapter, you don't need to go into much detail—this will come later in the Methodology chapter.

In research papers, the introduction is about identifying what is usually referred to as the "gap in the knowledge" that this paper is going to address, and it is usual to complete the introduction with what is known as the "research question". It is advisable to take a similar tack with your dissertation report. Finish off your introduction with a clear statement of the aim of the project, followed by the objectives you will target in order to achieve that aim.

Methodology

The Methodology chapter is the next main part. This is where you explain how you are going to carry out your project, and justify why you have chosen these methods, and not chosen other methods that might have been available to you. The word *methodology* technically means "study of methods", so merely stating what you did, doesn't necessarily make this a successful dissertation report.

Depending on the type of project, you might have mentioned some theory relevant to the project. In this chapter you need to set out in detail the underpinning theory involved in the project. In all likelihood, it will be possible to express the theory in the form of equations. This is the part where you demonstrate that you really understand the background you wrote about in the Introduction section.

If your project aim breaks down into a number of objectives, then you need to explain in this chapter what you did to meet each of those objectives. It is quite normal to break the chapter down into major sections for each objective, then to break down each of those major sections into minor sections to describe the tasks to be completed. The best reports will also draw attention to how tasks were scheduled using the Gantt chart at this point.

It is usual to record any information about your input data in this part of the report, clearly stating any assumptions made. It is also usual to include a note about any limitations you have imposed upon your project. These kinds of notes provide your markers of evidence that you truly understand your topic. For example, if you are carrying out a project comparing optical sensors, you might select to only compare six that are relatively inexpensive, commonly used and easily sourced. There are probably thousands of optical sensors available on the market, but it might take you weeks or months to track them all down and purchase them, which, in itself, would be a very expensive exercise.

Results

The results are a consequence of the tasks that have been carried out. Before you start the project you need to think about how you are going to record results and represent them in your report. Some results lend themselves to tables of data, and some are better dealt with as graphs. You might consider using infographics as a means of getting across your information; however, be careful that data in the report is clear, and not open to misinterpretation.

It is very important that you do not try to hide errors. In fact, it is recommended that you include a subsection in this chapter called something like "Consideration of Errors". By pointing out where you think you have outlier results, and making a recommendation of how they are dealt with, you are engendering the confidence of your reader in the quality of the rest of your dissertation.

The nature of your results will vary depending upon the nature of your project. The presentation of results should be factual and without interpretation—interpretation and analysis should be reserved for the next stage of the report.

Discussion

The word *dissertation* means "discussion", so this is the most important chapter of the report. This is where results must be analysed, and the analysis interpreted with respect to the background, or context, of the project. If it's appropriate to refer to some of the earlier published research in the topic of the project, then do so. This is known as "synthesis".

You need to plot your discussion carefully making sure you cover each of the objectives of the project that you set out in your introductory chapter. The poor dissertation report will contain objectives that are never addressed. Some students think that the report needs to be written like a diary, and the whole thing submitted— almost like a record of what happened, as it happened. Of course, you would have worked out that if your project isn't entirely going to plan, then change the plan. For example, you might have identified half-a-dozen objectives at the outset, but as time passes you realize that you are getting behind schedule in achieving each objective. This is when you need to find the "run-offs"—what tasks can you omit, but still achieve your objectives, or which tasks can you omit and achieve a coherent subset of your original objectives?

The other aspect of the discussion that is important is a critique of the processes that have been undertaken to complete project. In your methodology chapter you will have justified using particular methods; in this chapter, you need to demonstrate how the outcomes support those justifications.

Conclusions

The conclusions chapter is where you bring all the outcomes of the objectives together to show that the aim of the project has been met.

The conclusions must not introduce any new ideas. (Refer to Chap. 11.)

Recommendations

Suggestions or recommendations for further work are best dealt with separately to the conclusions of the dissertation. The obvious questions that are raised by notes in this chapter are: "why didn't you include that in your project?" So, in your defence, you should really support your recommendation with the reason why you omitted it. This might be a simple fact that you didn't have the time or resources to pursue a particular line of enquiry. (Refer to Chap. 12).

References

Be consistent with whichever of the systems of handling references you have selected to use. This is discussed elsewhere in this book. Refer to Appendix A.

Appendices

What you include in the appendices will depend upon the nature of your dissertation project. It is generally thought that appendices should only be used for additional information—there should be no repetition of what is in the main body of the report. Whatever additional information is included should only support understanding of the report. For example, if your project has been about the development of an app, then the appendix might contain code files to demonstrate evolution of the program.

14.1.5 Assessment of the Final Report

Each institution, and perhaps each engineering department, will have assessment criteria for the final report. Make sure you know what they are before you start writing-up your report. Sometimes, a detailed weighted breakdown will be made available to you; however, this is not always the case.

When supervisors get together to compare notes, generally everyone knows what a superb final report looks like, but it is challenging to define it categorically. The elusive star quality is probably best described as "coherent". The final report has a narrative arc that describes the story of the project. There will be many strands to the story, and these will be obvious in their continuity as themes in each chapter. The context of the project will be evident; assumptions will be clearly stated, and methods will be appropriate for the problem being considered. The analysis of results will support the aim of the project, and the conclusions will be clear and definite.

14.2   The Presentation

14.2.1   The Purpose of the Presentation

There is often some sort of live presentation for a project; although there are various formats this may take. Similarly, the purpose of the presentation may vary from course to course; so, whatever we advise here, make sure you know what the brief is for your particular course. It is also advisable to find out its relative importance; for example, in some institutions it might be a lightly weighted component of the overall assessment of the project (say 0–10% of the final mark), but be mandatory, so that it is impossible to complete the degree without attending this oral examination.

In some places the presentation may be called a "viva voce" (often abbreviated to viva). This is Latin for "with living voice" or "by word of mouth". The implication of this is that it is a live performance with the examiners present, all in real time, rather than a recording, although a viva might be recorded. Traditionally, all doctorate degrees in the English-speaking world are finally assessed by a viva. It is not universally the case that all undergraduate and Masters degrees include the requirement. In a world increasing concerned with authentic assessment, veracity of authorship and guarding against plagiarism, use of the viva for the more junior degrees is increasingly being used.

The purpose of the viva at undergraduate or Masters level is generally for the student to demonstrate that they are the author of the project (and its report), and that they understand the theory, which has been used to underpin the project (i.e. its application). A good presentation will contain an element of criticality—discussion of the decisions that have been made in carrying out the project, including a brief explanation of the reasoning behind the making of a particular decision.

14.2.2   The Audience for the Presentation

Different institutions will also have differing rules about who is required to be present, who can attend and to how many people the presentation must be made before. Common practice is usually consistent within a country. So, within the UK, an oral presentation will usually be treated in the same way as any other examination with perhaps two academics ("examiners") and perhaps a member of administrative staff (who is invigilating the exam process to ensure consistency in the way the exam is conducted). It might be the case that one of the academics is the student's supervisor, or, conversely, staff might have been selected as independent examiners. Another variation is whether the examiners have read the final report or not. Part of the assessment might be that the student is communicating their work to people who have no prior knowledge of it, so in those cases examiners may not yet have read the report.

In parts of Europe, the requirement is that the presentation is a lengthy "defence", which is read-out before a public audience. This is to ensure transparency of academic standards within the university.

In some places, the viva is made before an "invited" audience—invited by the university, a ticket-holders' affair—but this often means that friends and family can be in the audience (Fig. 14.2).

The duration of the presentation may vary. It is possible (especially where there is a large cohort of students) that each candidate has only 5–10 min to present an overview of their project, followed by a short period of answering questions, or the whole viva might be a series of questions-and-answers—more like an interview than a performance.

Fig. 14.2 Rehearsal improves performance (by Mr William Heath Robinson)

14.2.3 Planning

So, some questions to ask yourself:

- What format is the presentation going to take?
- Where is the venue?
- What time does it start? Do I have to register by a particular time?
- What is the duration? Does this include time for Q&A, or is that time extra?
- Who are the audience?
- Are supporting slides required or preferred? Is a computer provided, or should I take my own laptop?
- Can my presentation include videos?
- If I have artefacts, then may/should I take them along?
- At what point in the project does the presentation take place?

14.2.4 Presentation After Submission of the Final Report

There is no rule that the timing of the presentation has to be at the end of the project, but this tends to be the norm, similar to what happens in the case of the defence of a PhD thesis. This is a very academic approach to the presentation. If an oral examination of the project takes place at this stage, then the project is complete, and it is probably expected that you will report on all aspects of the project: its context, theoretical underpinning, methods used and their validation, results obtained and how errors were accounted for, an analysis of the results relative to the theory previously indicated, and a clear conclusion of the project. In addition to all the narrative of what was done, the examiners will probably be looking for an indication of what was learned. So, if there have been particularly important lessons learned— such as the time it took to learn a new software package, or to engage with your sponsor's procurement process—then this should be mentioned. A word of warning here; however, any issues that you faced and had to be addressed (which might have given rise to these lessons learned) and should not be presented in an overly negative manner. No one likes to sit through a presentation that is just a set of complaints, so try to keep negative emotions out of it.

14.2.5 Presentation Prior to Completion of the Final Report

In many ways this is more an authentic approach to the oral presentation exam. It corresponds to what happens to experienced researchers, who will often present their early work at a conference of peers in order to get questions or suggestions, which help to clarify or improve their work, or make it more relevant to the wider research community.

It is also more normal in industry to present progress on a development project before it is completed, so that a project client can see how things are progressing. This can lead to projects being extended, or new phases being added to existing commissions (so more fee-earning opportunities). The presentation is a means to developing a good working relationship with a client.

For the dissertation project, the presentation of work-in-progress can be daunting. It can raise a host of self-doubt as to whether you have done enough work, or whether you are going to get the dataset you need to complete your analysis. For the good student, who has been blessed with a good supervisor, self-doubt shouldn't be an issue, as you will have both monitored your project against your Gantt chart, and know whether you have used-up all the built-in slack yet! The worst-case scenario is possibly the confident student, convinced they have completed the assignment of the dissertation, to whom the supervisor suggests one additional thing that would really enhance the final report—something that might sound relatively trivial, but requires an entire chapter to be re-written. Take care that you understand the implications of any such changes.

14.2.6   *Presentations Other Than Live Performance*

Earlier in this chapter, it was said that the oral presentation was partly to test the authorship of the project, and hence the requirement for a live performance. It is also the case that engineers need to practice how to promote themselves and their projects. For graduates who want to go on with the careers as engineering scientists, there is a particular need to learn how to communicate and disseminate their research. Some ideas are discussed below:

Poster
Poster presentations are particularly popular with academics. There is a tradition amongst researchers to present at technical conferences. If your application to present a paper is not accepted, then the consolation prize is the offer to present a poster. The poster is a summary of your paper containing the essential information, without the need for you to be there in person to explain it, but if you were to attend, your verbal explanation would be supported by the graphics, figures and text. (Refer to Sect. 16.1.)

Video
We are living in the video era—if you want to learn about anything, there is probably a video freely available to give you instruction. So, unsurprisingly, some universities ask their students to prepare short video presentations. It can take HOURS to make a clip of a few seconds, and there are various ways of doing this. You might put a verbal commentary to your PowerPoint presentation, or you might make a talk-to-camera type performance. Whatever you do, be prepared to rehearse, and shoot many takes before you get the final take that you "publish". The one you make in your bedroom, with all your childhood cuddly toys in-shot on your bed, is probably not going to

impress the right people—much better to go for the sterile white wall behind you as you address someone (real or imaginary) just to one side, behind the camera. You don't have to publish your video on one of the big platforms, such as Instagram, YouTube or Vimeo, but you might find that is the most effective way of getting it seen, as you can then use a link to it from your other social media presence. It doesn't have to be of TEDTalk standard, but have a look at some of these for inspiration.

Website

Having your own website is the obvious means to publicizing your project (and self-promoting), and given the tools, like Wordpress, that are around now, you don't need to be a proficient web-designer to get something up-and-running relatively quickly. If you have to create a short presentation for your formal submission, then you can use the slides as the basis for content of the website for wider dissemination. Prospective employers are impressed to be able to view a clearly structured, well-composed online presence from someone they may be interested in employing.

Additionally, a website could be a good way of using any videos related to your project—this could be a CFD sequence that you form into a GIF, or testing your Formula Student car. Demonstrating performance success is something to celebrate!

An alternative tactic for using a website is to blog your project progress. This might sound somewhat tedious, but if you are doing a design-make-and-evaluate type of project, it can make a narrative full of drama and suspense. And it will give your elderly relatives the opportunity to keep-up with what you're doing at university!

App

Of course, not all projects are going to be suitable for making an app, but if yours is, then this is a very effective means of disseminating your work. Make sure to use your social media presence to promote it.

All of the above ideas can, and probably should, be used in combination, in the right circumstances.

14.3 Intellectual Property

14.3.1 Intellectual Property (IP) and the Dissertation

As a future professional engineer, you have hopefully been taught quite a lot about intellectual property, its nature, content, legal position and protection, whilst your supervisors and other collaborators as professional engineers themselves should already be aware of the main issues. This chapter isn't intended to substitute for that teaching and experience, or provide a comprehensive course on IP; there are whole books on that—and often they are country-specific. What we've tried to do here is show you the main things you need to think about in respect of any IP, used or created in the course of your project.

In the majority of dissertation projects, what you generate will be your intellectual property, while it most likely contains nothing of commercial value—it is after all a student project done using limited resources in a short period of time. But occasionally, a project does create something that *might* be of such value, or which the student wants to try and exploit—perhaps starting their own business after graduation. It's worth at least thinking about this possibility.

Around the world it's accepted that people who create IP should have the ability to protect and exploit that property. In the context of a dissertation this might be IP created by a student, IP already created by a project sponsor and used within the project, or details of a broader project that the project fits within. Designs, mechanisms, original text, photographs, ways of solving problems, ideas (even if still in somebody's head) and pieces of software (including an app) are all by definition intellectual property—the important questions are "who owns it?", "does it need protecting?", and if so, "how?"

14.3.2   Who Owns the Intellectual Property?

Usually, IP belongs to the person who created it. If they were being paid to create it, then *probably* it belongs to their employer—although hopefully they have an employment contract which clarifies that point. A university student is in an unusual position because they are most likely paying to be at their university, but nonetheless are subject to the university's rules—which often will lay down principles about IP ownership. The following is extracted from rules at the University of Cambridge, for example[1]:...

14. The entitlement to intellectual property rights in material created by a student shall rest with the student, with the following exceptions:

1. (*a*) Where a student is sponsored by a third party, a condition of sponsorship may be that the sponsor may own any intellectual property developed during the period of sponsorship. Sponsored students are, therefore, advised to check the terms of their sponsorship agreement.

2. (*b*) Where a student is working on a sponsored project as part of his or her course-work or research, the sponsor may own any intellectual property that the student develops. This will be specified in the research contract and the supervisor or Department should inform students if this is the case as early as possible in the admissions process and in any case prior to start of their research.

3. (*c*) Where a student is working in collaboration with others in a manner that gives rise to joint creation of intellectual property, or interdependent intellectual property, the student may be required to assign intellectual property to the University or place the results in the public domain without restriction. He or she will be treated in the same way as University staff under these regulations. If this case is likely to arise, students should be

---

[1]University of Cambridge, *Statutes and Ordinances*, Chap. 13, *Finance and Property*, accessed online 30 June 2019.

so informed at the offer of admission where practical, and in any case prior to the start of their research.

4. A student who believes that clause (*c*) above has been inappropriately applied may make an application to the University Technology Referee under Regulation 15.

A sponsorship agreement may also place a requirement on the student and his or her examiners to undertake to keep results confidential while steps are being taken to protect intellectual property or to establish exploitation arrangements. The student may also be required to submit the dissertation to the sponsor for scrutiny before submitting it for examination. Any confidentiality agreement whose purpose is to delay public disclosure for the purpose of protection should usually not have effect for longer than three months from the time the sponsor is notified of intent to publish. When the University obtains an assignment of student-created intellectual property, it undertakes to provide the student with a share in such financial returns from the exploitation as there may be on the same basis as that applying to University staff by virtue of Regulation 25.

Whilst Cambridge's rules provide an example of good practice, we can't possibly give guidance in this book on every university's rules, or even just what's normal for every country in which you could be studying. So, we can only strongly recommend that you have found, read and understood the rules at the institution where you are studying. Students who think that they have a problem are best advised to talk to their project supervisor first, but if things are particularly complex there should be a department (most likely called something like "research support office") who are experts in IP protection and exploitation and they should be happy to discuss issues and give advice to any member of the university, whether staff or students.

14.3.3   *Patents*

Patents are a contract between the state (and ultimately the world) and the creator of an idea or invention. They give the creator the exclusive rights to exploit or license their idea to someone else for a period of time, in exchange for the details of that idea being placed in the public domain so that it can be seen by anybody who wishes to search for it. Unfortunately, in most countries this exchange isn't free and the creator is likely to pay an initial and subsequent fee to do so.

An interesting historical example of this trade was in Britain in the 1920s and 1930s, where German company *Chiffriermaschinen Aktiengesellschaft (or Chi-MaAG)* had filed in Britain and America patents for the principles of an encryption system designed by its founder (electrical engineer Arthur Scherbius) that came to be known as *Enigma*.[2] A few years later British cryptographers were able to access the publicly filed patent information (Fig. 14.3) to understand how German military Enigma machines worked, and thus eventually were able to decrypt German military intelligence approaching and through World War II. This is obviously an extreme example, but it illustrates how the system works, and has done for many years (it did work both ways, British aeronautical engineer Arthur Whittle had filed patents in

---

[2]George Dyson. *Turing's Cathedral*. Pantheon (2012). Chap. 13.

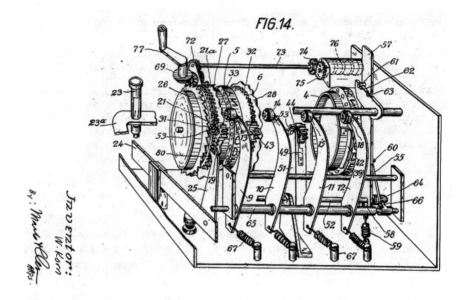

Fig. 14.3 Illustration of the working of an Enigma machine of 1923, from an American patent application that was granted in 1929

Germany for his jet engine—and almost certainly German engineers accessed those to support their own jet engine development).

Once a patent has been filed, access to the details are not normally restricted, only the ability to exploit it. So if, for example, a project client has filed a patent on an invention, and the dissertation is related to analysis of that invention—openly using that patent, and citing it as a reference in the report, should be absolutely fine. However, if a patent has not yet been filed but there is an intention to, then there's likely to be a need for considerable secrecy—any public knowledge of the idea of invention can prevent a successful patent application. If such secrecy is being demanded within a dissertation project, this could create major problems—as the ability for dissertation work to be read by internal and external examiners at the very least, is essential. This is too special a case to offer general advice: if something like this comes up, the best action is to get the key people together, check university rules and the patent process, and find a solution which is satisfactory to everybody. If you are a student who has come up with an idea, invention, process, and so on, you would like to try and patent—for example, with a view to starting a business after graduation with it, then you should probably ask for that discussion as early as possible, but keep it to very few people.

14.3.4   Copyright

Whoever creates an "original work" normally owns the copyright on it. A simple statement somewhere discrete on documents or media along the lines of "© A Student, 2021" is a traditional statement of this, and whilst not strictly required, a good idea if there's any risk of needing to assert this at any point. A simple and traditional way to ensure you can prove copyright is to place your work (in any form, printed, recorded, on media) in an envelope and post it to yourself. A postmark on a sealed envelope which can be opened, for example, in court by an independent legal professional, provides trusted evidence of the date on which it was yours. Of course, as with patents, this is not a necessary concern for most student projects—but in a few it may be, and we're mentioning it here as something which *can* be done.

14.3.5   Non-disclosure Agreements

Where a project is in collaboration with the owners of some intellectual property, it is quite common practice to be asked to sign a *Non-Disclosure Agreement* (NDA). This is a document which normally commits the signatories, or their employees, to only using whatever information they've been provided for a pre-stated purpose, not disclosing it to anybody else without permission, and returning all material on request.

Whilst NDAs are a common tool in business, they can cause a problem in universities where mechanisms may, or may not, exist for their signature and enforcement. It's not that unusual for individual engineering academics to sign them when participating in a consultancy project—but it's much more difficult for this to encompass students, technicians, examiners and so on. Where reasonably possible, it's best to avoid their use. If a project client asks for NDA signatures, this should probably be referred to a university's legal department and in some cases it may be necessary to decline the project if this can't be resolved.

Alternatively, if later the student or other members of the project team want to consult with specialists about exploiting it, asking people you consult with to sign an NDA might well be a reasonable precaution. Searching online will find many examples of NDAs you can use as a template, although check that what you propose to use is valid in the appropriate legal judisdiction(s). Of course, if a specialist declines to sign an NDA, you then have to choose whether to decline to meet, or change that requirement. We have tended to prefer relying upon our codes of practice as Chartered Engineers, but that is just our approach and may not work for everybody.

14.3.6   A General Caution About Intellectual Property and Its Protection

Protecting intellectual property is expensive, and can often be pointless. Even if legally and morally you have rights over the material, initially paying to file a patent, then fighting to protect your IP can be extremely expensive. Also sometimes, there really just isn't anything that bears protection. There is a significant risk that the only real winners in such a battle will be lawyers. Whilst there have certainly been cases where both individuals and companies have successfully and profitably protected their IP, there have been many more where money has been effectively thrown away for no real benefit.

Many inventors and product developers have taken the view that the best protection of their ideas is reasonable discretion, combined with ensuring that they understand their idea and its exploitation better than anybody else. These are, generally, much cheaper—and the second protection, understandably, is essential to persuade anybody to work with you on your idea—with, or without, IP protection. If you think that you have good ideas worth exploiting, these two principles are essentially good practice in every case: at that point it's worth **also** considering more formal IP protection, but it may well not actually be a good idea (or affordable). The one universal point to remember, however, is that if you're going to file a patent, do it quickly before your ideas become known to enough people to ruin the patent application process.

Chapter 15
The Dissertation and the Job

Abstract Most engineering graduates hope to progress either into a graduate job or higher study such as a PhD. The experience, and fact, of the dissertation provides valuable material and learning to support either of these, both in applications and when starting work. When pursuing a professional engineering job, the dissertation may have built up useful contacts, and provided enhanced knowledge of an industry sector or particular role. If applying for a PhD the dissertation also provides specific evidence of the ability to produce independent research and writing about a topic: a vital requirement of any prospective PhD research student. In commencing a new role, used well, the learning from the dissertation can also give a head start in skill and knowledge.

© Springer Nature Switzerland AG 2020   203
P. Gratton and G. Gratton, *Achieving Success with the Engineering Dissertation*,
https://doi.org/10.1007/978-3-030-33192-4_15

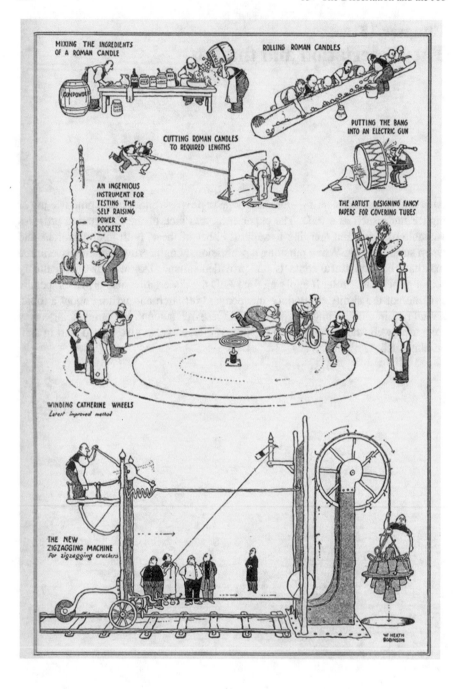

15.1 The Dissertation and the Job Application

15.1.1 Introduction to Job Applications

"Success" is something that can be reckoned in many ways. One specific form of success is to get your dream job off the back of what you do for your dissertation project. So, this chapter is about how you can use your engineering dissertation to get a job, or more generally, getting into the next phase of your career as a professional engineer, whether that is in employment or as a research student.

This chapter is about using your dissertation to help you make a successful job application. Far too many students and graduates, after they have made dozens, if not hundreds, of job applications, still have not got anywhere. What they're usually doing is looking on the recruitment websites and sending applications to every vaguely interesting post with little consideration for customizing their "standard" resumé. This is like planting a crop of rice by discharging seed from an aeroplane at 500 feet. There is an element of risk that all the seed you drop won't hit the fertile soil, or, alternatively, that plants grow in such a way and in places that any viable crop is going to be impossible to harvest easily. The approach that's more like to yield a return is "targeted": plant the right seed where it is most likely to grow.

Ahead of applying for a job you need to decide where to apply and, more importantly, what kind of job you think would like to do, and would be good at. Assuming you have managed to find a topic for your dissertation that you also enjoy and are good at, you can use this as a means of finding a role that will pay you to do what you enjoy. For example, if your dissertation is about atrium buildings and the effect on energy use in buildings due to atria, then perhaps you would like to work in designing atrium buildings—perhaps in architectural engineering or building services design. To find out where the companies are that do this kind of design, it doesn't take long to use a search engine to look up "recently completed atrium buildings" (there were millions when I did this). If you have access, via your university library, to one of the business analyses databases, such as ORISIS, or to Nexis for newspapers, then you might find a more useful search of professional media—the trade press and publications. You are more likely to find informative articles about atrium buildings in the Architectural Association journal than through Google News.

15.1.2 Before You Start Making Applications Make Some Decisions

Before you go looking for a job to which you want to apply, there are some practical considerations to take into account. Let's classify these as *constraints* and *desires*. The lists of constraints and desires suggested here are not exhaustive but are the commonest ones we could think of. Also, you are likely to weigh each one differently depending upon the circumstances of your life at the point of making a job application.

What are your constraints?

Geography can be both a constraint and a desire. Many students have "gone away" to university on the understanding that they will "go home" once their studies are complete. So, to progress their career, they specifically look for jobs near home—or at least within an easy commute of home. Alternatively, a graduate might now be in a relationship with a partner about to start a PhD somewhere new. So, they might decide to look for a job in that city for a few years with a view to reviewing where they want to live after their partner's research degree is complete. Setting a time-constraint on that first job is treating the job search in industry rather like selecting a dissertation project. Being candid with the recruitment consultant can pay off.

A much more important constraint is about whether the job, or the company, aligns with your personal ethics. We all have values which affect our opinions and behaviour. The law of the land and societal conventions require that we control our behaviour, and how we express our opinions. There is little more stressful than having to work in an environment that constrains, or challenges, our personal values every working day. So, heed the advice of the Delphic Oracle and "know thyself!" If the idea of such devices and their use is abhorrent to you, then working for a company that make anti-personnel mines is not for you.

Evaluate your dissertation project in terms of your personal values—asking yourself what attracted you to the topic; what is it about the topic that keeps you engaged—these might be the characteristics that you need to thrive in a job. You might not know what job you want to apply for, but if you can identify the organizations that are advertising in terms of their values, then you are more likely to find one with values that overlap with yours.

What are your desires (e.g. experience, travel, money)?

We carried out a straw-poll with a group of graduating students asking what they hoped to gain from their first job after graduation. The most popular response was *experience*, and when asked to explain, the consensus was that it was a good idea to get a well-known employer on your CV early in your career, because that would "open doors" at some later date. This may or may not be true. Sometimes the kind of useful experience is to be gained in the jobs in the smaller companies, or those who are part of the supply chain for the more famous brands. Future employers are more likely to be interested in the experience gained and the types of projects worked on, starting with your dissertation, than that you spent time at one company or another.

The skills you have learned and developed on an engineering degree are broad and transferable. Don't feel you need to pursue jobs related to your specific dissertation content. The dissertation gives you an opportunity to develop project management skills that are valued by employers worldwide.

If you are interested in commercialization of your project, then there is likely to be an "enterprise" office, or commercialization support office within your university—they specialize in putting people with ideas in contact with financial backers. You'll need to be clear about what you want and what you are willing to trade.

15.1.3 Talk About Your Project—Including Friends and Family

A simple way of using your dissertation to help find a job you want is to talk about your project to anyone who'll listen. To start with, talk to friends and family. Hopefully, your friends, particularly if you know them through student and professional societies, will be interested in some of the things you are interested in, and talk freely about the ups and downs you're experiencing as you carry out your project. Family often have a vested interest in your success, and want to know about what you are involved in, so make a sympathetic audience. It is possible that neither your friends nor family have your technical education, so you might find that you need to explain your project in non-technical language. This is excellent practice, as it is exactly what your future employer (and examiners) are looking for—clear, concise explanation of your work. The reason for talking to friends and family are:

- They're a friendly audience to practice on
- You never know who they might talk to about your project, and where that might lead!

Some years ago, one of us asked a former student how he got his job building controlled environment cabinets for the pharmaceutical industry. He was bemused to explain that his mother had "found" the job, whilst she was at a beauty spa. As she was enjoying a treatment, she fell into conversation with another customer whose husband was an industrial chemist. He'd successfully developed some new processes, which many bio-tech companies wanted, but they needed very accurately controlled conditions. He had a burgeoning order book, but only one engineer developing the specialist cabinets. "That's interesting," said Mum, "my son's just about to graduate in engineering. He's looking for a job, and talks about noxious emissions!" Phone numbers were exchanged and the rest is history!

15.1.4 Be Prepared to Talk to Random Strangers

Of course, what you really need is to condense your introduction into an "elevator pitch"—this is a succinct verbal presentation of what your project is about that you can deliver in the time between the lift doors closing at ground floor and getting to the "executive" floor. This is about 30 seconds to a minute. Of course, that duration is going to vary depending upon the transport system technology and height of building, but the point is there is only time to deliver a very condensed message. So, the information needs to be condensed to the essentials, and delivered in such a way that your audience says, "Tell me more!"

This happened to a former student of ours, Ollie, when he was staying with an old school-friend during the Christmas holidays. The friend's family had a ski-chalet where he was invited to join them. One day he bumped into the occupier of the neighbouring chalet, who whilst chatting asked him what he did. When he said he was an engineering student working on an entry to the Formula Student competition, the man's face lit up, and he explained that several of the people at work had been involved in that competition. He was technical director of a well-known sports-car manufacturer. Ollie got shortlisted for interview for a one-year placement, and upon successful completion, he was invited to join the company's graduate programme.

One further note—talk to people, but not too much or too often—everyone gets fed up!

15.1.5 Find Out Where the Specialists Are and Talk to Them

Of course, you can accumulate hours talking to friend, family and random strangers and never chance upon anyone who has a job, or knows someone who has a job. This time is an investment in perfecting your personal pitch. If you are going to use your dissertation project to get a job, you will need to talk to the people who know about your topic, and are likely to know who is employing in the field of your topic. As you start out in the process of getting a job, it is probable that you won't know about all the types of roles that you could apply for; that would make great use of your knowledge and skill set. You can spend hours (even days) scouring the "vacancies" websites, but the adverts are often written in a kind of language that lacks promises, or any hope of being interesting. Nothing quite beats finding someone who knows about the jobs in the industry where they work.

Before you go tracking down a Special Interest Group (SIG) meetings in a city near you, you might search the internet for specialist bulletin-boards. Online communities proliferate! For every official forum there is likely to be an unofficial one. It is worth remembering that the practical limitations of digital communication tend to make discord inevitable. Chatting in cyberspace is not the same as talking in real life, so find out where groups that might be useful to you meet, and make the added effort of putting on a clean shirt!

15.1.6 Present About Your Project—Present Around the World (PATW)

The obvious communities of expertise that your project might intersect with are likely to be found in the professional engineering institutions and associations. These are all likely to be welcoming to nascent engineers—it's a way for them to gain new members—and many are likely to have competitions for new graduates. We discuss national and international competitions later in Appendix C.

15.1.7 Write About Your Project

The obvious place to write about your project is in your application, whenever you are asked to give examples of how you have applied your knowledge or skills. These days fewer job adverts ask you to supply a CV or resumé, more often filling in standard forms, but it is still good practice to have a version of your CV in a document, where you can keep information for copying into online applications.

Use social media to practice writing about your dissertation topic. LinkedIn is an obvious platform on which to advertise your project in the UK and USA, but other countries, and some disciplines, may favour other fora. You might also think about blogging about your project on a monthly basis. You ought to be having regular meetings with your supervisor on about this level of frequency, so blogging about the project might give you an informal means to crystallize what you discuss with your supervisor.

As we're sure you are already aware, recruiters use bots to scour the internet to find out about you. This is a fairly low-cost means of doing background checks on you. You already know not to use profane language online, but be careful also on what you say about your project, as it provides apparent insights into your nature which you need to manage. If you have waited weeks to get an order for components placed, and when you ask the technician about progress, she tells you that you submitted the wrong form, then do not use the blog as an opportunity to publically whine and moan about the technical staff, or the department. This will not make you look like a good team player. You might be in the "right" and the staff in the "wrong", but you are displaying a level of disloyalty likely to be unattractive to a prospective employer.

Finally, a simple test, assuming you are enjoying your dissertation project, is to ask yourself whether any job you are considering applying for is likely to be as fun as the project. If yes, go for it!

15.2 The Dissertation and the Job Interview

15.2.1 Introduction

In the previous section we emphasized the need to talk about your project, so that you can talk about it fluently, authoritatively and with confidence. Of course, if you are successful in making a job application and get called for interview, then all that talking about your project will have been good practice. However, before you go to the interview, there is more preparation to be done.

First, you need to review everything about the job application. Re-read the advert, job description, person specification and, of course, your own application. Secondly, with specific respect to your dissertation project, think about how it relates to the job description and person specification—are there any specific points in the description that align with the topic of your project. For example, if the job is about battery-powered electric vehicles, and your project was about optimization of battery performance for application in an electric vehicle, then you need to think of all the things you learned in carrying out the project.

15.2.2   Interview Technique

Talk about your project with respect to what you learned from it
Elsewhere in this book, we talked about the need for reflection, and described Gibb's model. As you prepare for interview, use the model to talk about your project. An interviewer is likely to be more interested in what you found exciting or disappointing; what you think was good or bad about the project; what you learned from your dissertation; and how you think this might help you in working for their organization.

Of course, dwelling on problems encountered during the project might convey too much negativity, but interviewers are going to be interested in problems overcome, and perhaps less likely to want to hear the details of the methodology.

It is, of course, still quite possible that the job you apply for is not very related to the project you undertook for your dissertation, but there will be lessons learned that are transferable to other types of projects in other situations. So, think hard about the job applied for, and what lessons are transferable to it. This addresses one of the top five interview questions—"what do you think you bring to this role?"

Tackling competency-based interviews
A very common interview technique focuses on getting a job candidate to explain how they work, or deal with people, or tackle certain situations. The interviewer is looking for specific competencies. It will help you to prepare for these types of questions by using the STAR mnemonic:

What was the **Situation** that you are considering?

What **Task** needed to be done?

What **Action** did you take?

What was the **Result** or outcome of your action?

The competencies the recruiter is looking for will usually be described in the person spec for the post. You just need to think of a STAR routine from your experience, drawing on your dissertation project that satisfies each competence listed.

The multi-part selection process
We have only focused on interviews here—because it is an opportunity to use your experience in the project to promote yourself favourably to a prospective employer. It is quite possible that an interview is part of a much larger selection event. One engineering student recently told one of us about an assessment centre he attended that took a whole day, including a presentation from one of the company's top engineers, psychometric tests, a group exercise, an interview with an HR manager, and one with an engineering manager and a tour of facilities. A total of 200 people started the day. By the end of the process, only 20 were selected for the tour, and they were

informed that only 12 would be subsequently contacted with an offer for a place on the graduate scheme. He was one!

Many companies are using online psychometric tests, and video interviews, as an alternative to the full-day assessment centre. However, it is quite usual for interviews to be at least two parts—one with a member of the HR department, and the other with the line-manager for the post. The HR officer is usually making an evaluation for the characteristics set out in the job description. One of these is likely to be how well you communicate with non-engineers. Be careful, however, not to oversimplify any explanation of your project—this person works with engineers on a daily basis.

Only your dissertation project?
Whilst you can use your dissertation project as a relevant focus to a job interview, you will need to demonstrate to a prospective employer that you are a rounded individual. So, although you have spent an intense period in the last year working on your dissertation, it should NOT be the only thing you talk about. If it is, then you are not likely to be successful—a balance has to be struck!

15.2.3 For More Advise About Interview Questions

There are lots of books available on preparing for job interviews. John Davis writes specifically for engineers, and has a chapter in his book on interviews.

Davis, J. W., *Communication Skills: A Guide for Engineering and Applied Science Students* 3rd ed. Prentice Hall (2010) pp. 158–164

For a more general book that discusses a list of interview questions in depth

Lees, J. *Knockout interview*, McGraw-Hill Education, 4th ed. (2017)

Many websites also provide advice. An online article aimed at engineers is:

Lloyd, J., *Job interview tips for engineers*, CV Library Ltd. (2018) Available at https://www.engineeringjobs.co.uk/career-advice/featured/interview-guide-for-engineers/ [Accessed Aug 15 2019]

One based in the UK, not specifically about interview for engineering jobs is

Monster Worldwide, Job interview questions: the ultimate guide (2019) Available at https://www.monster.co.uk/career-advice/article/what-are-the-most-common-job-interview-questions [Accessed Aug 15 2019]

15.3 The Dissertation and the PhD Application

15.3.1 Introduction

For some students, or graduates either straight after graduating or a few years into their working lives, the next thing they want to do is a PhD. Perhaps not many—after all most people go into engineering to become engineers, rather than to become researchers. But, some of us either as students, or later, decide that we've got a fascination with either a particular subject, or with the research process, and want to put ourselves forward for a PhD. The PhD is referred to as a "terminal degree"—the highest level of university qualification you can obtain, but in reality it's a research licence and the PhD is all about how to conduct independent research. The actual programme will vary between institutions and countries—but ultimately it's about producing a large thesis—typically six times the length of an undergraduate dissertation or four times the length of a Masters dissertation, which demonstrates a genuine contribution to human knowledge, and a mastery of some small sub-section of that knowledge. To supervise a PhD student is an enormous investment, by an academic, their department and the university—so they are clearly going to be extremely selective of who they take on.

15.3.2   Demonstrating Your Potential as a Researcher

Whilst it was on a much smaller scale—your Bachelors or Masters dissertation will help demonstrate whether you have the potential to be a successful PhD student. All of the basic skills required—competence in an engineering discipline, researching from a variety of sources, managing a project and technical or scientific writing will be there. So, it's highly likely that a prospective supervisor will either want to see your dissertation, or test you on something very similar (such as writing a PhD research proposal) before accepting you. It's also possible that they'll favour applicants whose dissertations were in topics similar to the planned PhD research. [Note by the authors—this is not a universal view, so don't lose sleep on this last point.]

For a student who wants to use their dissertation to help them get onto a PhD programme, first they need to produce a really first-class dissertation, consistently styled, well structured, beautifully and consistently referenced, and where every recommendation is built on a conclusion, and every conclusion built on a clear narrative flow of evidence and argument. A good mark for that dissertation is obviously a bonus, but any good prospective PhD supervisor will read the work themselves and form their own opinions: so *absolute quality* will matter more than meeting local preferences and "just" getting good grades in your home department. Clearly, also make sure you have a printable electronic copy of your dissertation available to provide to any university you're applying to do a PhD with.

You must expect to be quizzed on, maybe even be asked to give a presentation about, your dissertation during a PhD interview. This means that you can't afford to file it away and forget all about it after finals: you want to arrive at a PhD interview knowing your way around your dissertation, and being able to describe the work you did, how it all went and what you learned from it. Even if the PhD you want to do is in a completely different subject, a prospective supervisor wants to know that you can research, solve novel problems and explain your thinking and actions—that is usually far more important to them than your narrow subject knowledge, which they can find out from your exam grades anyhow, and will invariably need to build upon during your PhD.

15.3.3 Getting a Reference for a PhD Application

You should also bear in mind that all the people you came in contact with during your dissertation—your supervisor, client, lab technicians—will all potentially be referees for you as a prospective PhD student. Whilst you are likely to be asked to provide the PhD university with two or three referees, and of course you would have carefully selected them to be the people who'll give you the best references, reality is somewhat different. It's very easy to see where you studied, who taught you, which lab you did your dissertation in and so on. Of course, this "connectedness" goes on across academia and industry, so it's not unlikely that something similar may also happen if you are applying for a graduate post in industry, joining the armed services or civil service, and so on, but the network in academia is particularly strong, and this becomes the most critical. So, all of the advice throughout this book becomes *even more* important. If you got good grades, but regularly missed meetings with your supervisor or messed about and annoyed the lab technicians, then that might well get back to the university you're applying to do a PhD at: and you'll never know, but it'll be a bad mark against your name. Of course, the opposite is also true: one of us picked a PhD student once, who had fairly average grades, very much on the basis of glowing personal comments by their MSc dissertation supervisor at another university—and that turned out to be an extremely successful PhD. Remember—a PhD supervisor usually has applications from many applicants with excellent grades, but to them this is much less important than your overall profile as a scholar, and what they can find out about the dissertation and from the people who saw you do it, is often the best and easiest evidence they have there.

If you are lucky enough to get onto a PhD programme, then you will initially be building upon the skills you learned during your dissertation. The research component of a PhD will usually commence with some combination of a project proposal and a literature review—both things you learned to do in your dissertation. So it is also these skills that will help you establish well on your PhD. Hence what we said back at the start of this book—good grades matter: but should be treated on an equal footing with what you learn, and how it sets you up for the next stage in your career.

15.4   The Dissertation as You Start Your First Job

15.4.1   What Has Your Dissertation Got to Do with Starting Work?

The dissertation is the first taste for many people of what it's like to work as a professional engineer. Of course, it was artificial in many ways, since as a student you probably had much more autonomy in how you worked, and also that the constraints of time and budget were tighter and more rigid than would be normal in industry (although the constraints on labour time probably more flexible). But all of the fundamentals were there, and hopefully you've been hired into your first professional job, precisely because you have an engineering degree: whether that's in engineering, or in some other profession that values the same skills.

15.4.2   Transferrable Skills

Consider these skills, which you should have exercised and developed in the course of your dissertation.

Literature search
Starting any project, whether in university, in industry or elsewhere—for example, getting something up and running for a charity needs you to start by understanding what the state of knowledge is about the project's subject. You'll actually find this much tougher outside of a university than you ever did as a student for several reasons: you are unlikely to have access to university standard library facilities any longer, and you may often find that corporate culture understands the importance of literature search far less clearly than academic culture. Another issue is that almost certainly you'll find that however good your degree education was, once outside the university, there'll be much more to learn about—such as sector-specific techniques, technology you weren't taught about, your company's ways of doing things or even just brushing up stuff you did in the first year and discover you need again. So all engineers, especially those new to any workplace, will find it essential to spend a lot of time searching out and digesting new material.

Time planning
Working as a new graduate professional is likely to be very different to any student jobs you've had before, in that your line-manager is most likely to want to set broad objectives, and leave you to manage the detail of how you spend your time achieving those objectives. In your dissertation and, of course, in much of the study you've done up to it, you'll have mostly had to manage your own time, with only lectures, labs and meetings providing fixed points in your calendar. Take those skills to the workplace, and also expect that unlike life as a student, your employer is likely to also want to keep track on your time, so be prepared to keep detailed records.

Communication skills

A lot of your life as a graduate engineer will be about communication. Expect to be writing formal emails and reports, attending meetings and before too long, giving presentations (sadly, as the most junior person in the office, also probably writing the minutes quite often). This is very much the same as writing your draft and final dissertation, contacting your project client with updates on your dissertation or giving the end of project presentation—except that you may be doing this more and more frequently. If you paid attention to developing and maintaining these skills during your time as a student, this will serve you well during your time as a graduate.

Multi-disciplinary skills

You'll probably be relieved to know that virtually no job as a professional engineer requires you to regularly pass examinations in narrow subjects, like you had to do at university. Most engineers will find that they need to use multiple subjects at once, all at a high level and interrelate them to each other. This, of course, is exactly what you were doing in the analytical parts of your dissertation. It does come as a shock to many, but it's much better that you dealt with that unfamiliarity as a student, and not in the workplace where you're constantly being judged as a professional.

15.4.3   Subject Knowledge

In the dissertation project, you have spent a period of time—likely the equivalent of several months full time, working in the narrow sub-field of the project. If you picked your project (or got lucky) in something close to your graduate job, then there is a good chance you can use this knowledge as you start in employment to progress well in the early period. Quite possibly that study period will have given you a substantial advantage in the early months in your job—when of course your early reputation and working relationships are being established.

As you start in employment, you'll almost certainly be surrounded by engineers who have worked in the field for not months, but years and sometimes decades. So, the initial learning during the dissertation may give you a head start, which is useful to exploit—but continued learning from the first day in the job, particularly from experienced engineers with years of experience, will be essential.

15.4.4   In Summary

Hopefully, you have read this chapter for the first time before you've started your dissertation. Picking a dissertation project and using the dissertation *also* to develop valuable transferrable skills and sub-field knowledge can massively improve your chances of both getting a graduate job you want, and thriving in the critical first few months in that job.

All students should be thinking in terms of their subsequent career—as failing to plan and prepare means, in all likelihood, being pushed into the career route that was available (and other people didn't want) rather than something you do want. The dissertation, as the first proper taste for most students of professional practice, is a unique opportunity for career development. The wise student will grab this opportunity, from before they start, until they're sufficiently established into their early career that the dissertation has become unimportant enough not to list on their CV or resumé.

Chapter 16
Publishing Onwards

Abstract Some dissertation work may be suitable for publishing onwards. Several ways to do so are described: research posters, conference presentations, and papers in academic journals. Each of these routes are described, including how to approach these possibilities and the potential for success. It's noted that relatively little work from taught degrees will ever be suitable for publication, particularly in academic journals, but this can sometimes occur—usually in collaboration with a co-ordinating established academic.

© Springer Nature Switzerland AG 2020
P. Gratton and G. Gratton, *Achieving Success with the Engineering Dissertation*,
https://doi.org/10.1007/978-3-030-33192-4_16

THE NEW CRATER CRANE ERECTED
BY THE VESUVIUS GOLF CLUB FOR
THE RECOVERY OF LOST BALLS

16.1 Research Posters

16.1.1 What's a Research Poster?

Research posters seem to have started to become important around the 1990s when conferences increasingly wanted to give everybody who applied a chance to say their piece, but couldn't reasonably fit everybody into an already crammed lecture programme. So, the concept was created of the "poster session" where posters—usually around A1 (594 × 841 mm or 23.39 × 33.11 inches) or A0 in size (841 × 1189 mm or 33.11 × 46.81 inches)—are displayed in a room. Typically during poster sessions, at a defined time, their authors go and stand near their posters, and conference participants can walk amongst them, reading the posters, and have the opportunity to discuss the poster's contents with the author. Whilst they are often regarded as less prestigious than giving a formal talk, they do provide an opportunity for many more conference participants to present their work, and for much more detailed discussion with interested attendees than would be the case normally. The practice has spread so that often posters are used as part of dissertation assessments, student research conferences within universities and so on. So, you might have reasons to offer a poster (or be required to prepare a poster) in a number of contexts (including, but not only, at a conference—see later in this chapter).

Posters should be readable by somebody with average eyesight from about three metres or ten feet away. They usually only provide the key points of the work—whilst the temptation is always there to just see how much information can be crammed into a poster, nobody will thank you for that. Around 300–800 words, some good clear illustrations and an unmistakable message—that is what a research poster wants. Equations, technical diagrams and (IF they suit the message) paper-like constructions of introduction–method–results–conclusions are all fine, but they need to be bold and simple.

16.1.2 How to Prepare a Research Poster

Research posters are invariably created on a computer, and numerous packages are available, including those normally used for presentations, for desktop publishing, or just a word-processor. Templates abound for creating posters—you may find that one is provided for you by your university or a conference, or if you search online you'll find many hundreds of freely downloadable versions [don't forget to check them for viruses when you do download]. So, you can afford to work in software you're comfortable with, and normally start with a template at the right size. That said, sketching out your ideas first, just in pen or pencil on some large pieces of paper can often work well for your designing process, before going to the computer.

It's worth framing the image first with the things that have to be there—the title (use a sans-serif font, make it clear and easily seen—usually at the top), your name(s) as authors and most likely the logo and/or name of your institution. For a lot of research posters, this does get complex with a lot of names and logos, but for a poster from your dissertation, that's unlikely to be much of a problem. You probably need a simple title, your name or the names of your team and the logo of the university—perhaps that of the project client as well.

You might then want to start with your conclusion—where are you going to say it (either along the bottom, or in the bottom right corner are most common) and what is the big message you want to present. Then work backwards from there to make sure that, clearly and visibly, your poster contains a visual narrative towards the conclusion.

A good working rule is that if your poster is not readable printed out at A4 (approximately foolscap), or in extremis A3 (approximately ledger) then it's been over-done, and needs both simplifying and the size of text and diagrams increasing.

16.1.3   Example of a Good Poster

This poster in Fig. 16.1 we've not actually used (yet), and it's not perfect, but was designed, based around the project of creating this book. We've tried to follow the advice we give you here, including:

– Clear large and consistent fonts
– A clear heading
– Sharp, easily understood figures
– Symmetry that is easy to follow with the eye
– Good use of the space available
– Simple, but understandable information.
– Key stakeholders are mentioned, but not excessively.

16.1.4   Example of a Bad Poster

Conversely, Fig. 16.2 is a deliberate example of how not to create a research or project poster. As a learning exercise, have a look at this, and write down what you think are all the faults in the poster. At the end of this section, you'll find a list of the deliberate faults (if you found some more, well done!).

16.1.5   What to Print Your Poster On?

If a poster is going to be used in a session at your university—most likely only for a day or two—then just get it printed on plain paper, stick it up for the session, then it

Writing: *Achieving Success with the Engineering Dissertation*

Brunel University London

Petra Gratton [1] & Guy Gratton [2]
[1] Brunel University London, [2] Cranfield University

Springer

Cranfield University

The Concept

Create a book, useable by students, supervisors and project clients establishing best practice in student engineering dissertations.

Unique selling points

- Roles of the Student, Supervisor and Client
- Integration of the dissertation into career development.
- Appropriate humorous artwork
- Combining competitions with dissertations
- Engineering dissertations only.

Publication

- Hardback (online retailers)
- University library engineering packages from Springer
- eBook

Contact us about

- On-campus seminars
- Additional material
- Corrections / future edition improvements
- Advice on using the book in your courses

Supported by

The William Heath Robinson Trust and Museum, Pinner.

Springer-Nature

Contents

Introduction

What is the dissertation?	Purpose and structure	Types of project	The project client	Whose project is it?	Dissertation terminology
Objectives and expectations	Of the student	Of the supervisor	Of the project client	Of other stakeholders	

How to be the student who aces their dissertation

Selecting the project	Advice to students	Advice to supervisors	Advice to project clients	
The project plan	Time	Resources	Budget	
Designing...	Feasibility studies	Experiments	Products	Surveys

Acting professionally

Presenting data	Quantitative data	Qualitative data

Engineering analysis

Drawing and presenting conclusions

Making recommendations

Reflection

The output	The final report	The presentation	Intellectual property
The dissertation and the job	The application	The interview	The PhD	Starting work
Publishing onwards	Posters	Conferences	Journals and books	
Appendices	Referencing, citing, quoting	Using engineering standards	Design, make, evaluate	Competitions

Contact

ASED@gratton.aero

@ASEDtweets

Bibliography

1. JW Davies, Skills for Engineering and Built Environment Students: University to Career, Palgrave 2016
2. T Day, Palgrave Study Skills: Success in Academic Writing, Palgrave Macmillan Education 2nd edition 2018
3. C Winstanley, Writing a Dissertation for Dummies, Wiley 2009
4. SG Naum, Dissertation Research and Writing for Construction Students, Butterworth-Heinemann 2nd edition 2007

Fig. 16.1 Example of a good research or project poster

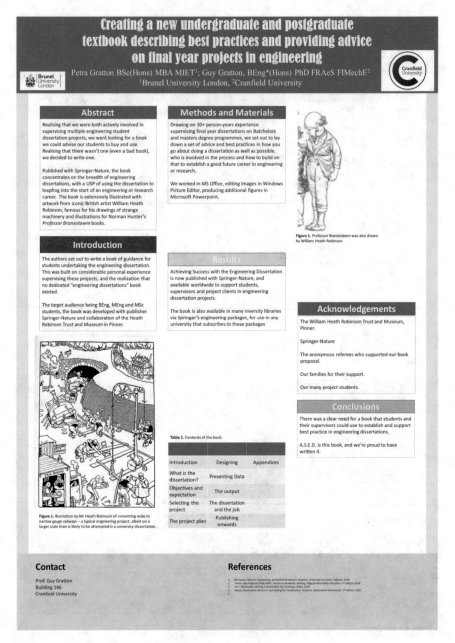

Fig. 16.2 Example of a badly produced poster

can be disposed of with minimal waste afterwards. If it's going to be at an important conference, but travel arrangements are easy, then the norm is to get it printed on stiff glossy paper and transport it rolled inside a cardboard or plastic tube. More expensive, but if travel arrangements are more difficult (say hand-baggage only on an economy airline) then printing onto fabric can work very well, as this is easy to transport and doesn't crease badly.

"Banner printing" on multiple pages, then sticking them together seldom works well, should be avoided if you reasonably can. More expensive mounting methods may be appropriate if a poster is going to go on semi-permanent display within your department.

16.1.6 At the Poster Session

If you are required at a poster session to be with your poster, then obviously be on time (which means five minutes early!), reasonably smartly dressed and fully up to speed on any difficult questions anybody might ask you about your work. In those regards, it's basically the same as any other conference. If the context in which you're presenting a poster is one where you're particularly trying to impress—networking towards a possible graduate job or future research collaborations, for example—then a small holder underneath the poster with your personal business cards and/or a smaller (say A4 or foolscap) copy of your poster may well be a good idea—unless explicitly prohibited by the event, nobody will ever criticize you for doing this.

16.1.7 Faults in the Bad Poster

- Small, hard to read text
- Large, unnecessary areas of space
- Multiple fonts
- A barely comprehensible heading
- Logos at different sizes (and two obvious logos of key stakeholders not shown)
- One figure too complex to follow along with over-long label
- Another label, blown up from something too small so not really readable (nor all that relevant)
- Random bright colours
- A bibliography incorrectly called references (but not called out from the text) inconsistently written, and too small to easily read anyhow
- Contact details that are essentially useless
- Shoehorning text under standard headings that really don't fit
- Repetitive information
- Excessive and unnecessary postnominals

– A table that contains potentially useful information, but in a form that's very hard
  to interpret
– Spelling mistakes (did you spot "niversity"?)
– Sometimes very long hard to read sentences
– Unexplained acronyms
– Does anybody really care what office package we were writing in? Say something
  meaningful!

If you spotted all of those—congratulations, you hopefully won't make similar
mistakes on your own posters.

16.2 Submitting and Presenting Conference Papers

16.2.1 What is a Conference?

Conferences occur in almost all fields of human endeavour, not least engineering and
engineering science. They are organized mostly by either commercial organizations
or by learned societies and professional institutions,many have sections dedicated to
student work, although not all (nor, always, does this matter). The format is usually
the same—groups of professionals, academics and some students meet in the same
place: typically a university building, institution or hotel for 1 to 4 days. There will
be one or more tracks of lecture programme, and in some cases discussion fora,
trade exhibition areas and/or poster sessions (see the previous section concerning
posters). Usually, but not always, admission requires payment, the amount depending
upon attendees status (age, employment, society membership and whether they're
presenting or not). Particularly for multi-day conferences, it's quite normal for there
also to be organized or ad hoc social events as well, whether that's meals, "cocktail
party" type events or technical visits to interesting sites. Talks are usually organized
in sessions of 3–6 talks between breaks, each session chaired by somebody eminent
(or just available!) from within the sub-sub-field being covered in that session.

The value of conferences is that they allow latest work and findings to be shared
with likeminded colleagues, and for those colleagues to network with each other. In
engineering alone, around the world there are thousands of well-attended conferences
each year—so clearly we all find value in them.

Admission to the vast majority of conferences normally only requires an application with payment. To speak or present at a conference, however, usually requires
you, in good time, to submit at the very least a title and abstract, which will then
be assessed by a selection panel who will consider whether the proposed "paper"
fits the conference, is likely to interest enough people and the proposed speaker is
authoritative on the subject. It is certainly appropriate for students to offer papers to
conferences, and very often there are both dedicated sessions for student work, and
heavily discounted or free admission for students and/or speakers.

16.2.2 What Exactly is a Conference Paper?

Historically, conference papers were usually very similar in format to journal papers—usually just shorter and less polished. This is still the case at some of the most important professional and academic conferences, but increasingly not so. For the majority of conferences now talks are around 20–30 min long, with an abstract required but not a more formal written paper. The talk will use a presentation method such as *Microsoft PowerPoint*, and probably be followed by around five minutes of questions. Conference proceedings are most likely, nowadays, to consist of the presentation slides and/or a podcast of the actual talk with the slides.

16.2.3 Applying to Present at a Conference

Usually conferences will have a website where you can download details of exactly what is required to apply to present and a deadline. The requirements are usually not negotiable, although the deadline often will slip at least once as conferences don't always get as many good papers by the first deadline as they hope for (but don't count on that!).

There's no value, usually, in doing more than the minimum that's required at this stage—typically an abstract and some personal details in your application—your presentation and/or paper can then be written once the paper has been accepted. BUT, particularly as a student or recent graduate do check what's available by way of funding. It is quite common that university departments, student or learned societies or organizations such as the Royal Academy of Engineering (RAEng) may have funds available that you can apply for to attend conferences as a new researcher presenting a paper. These funds are there to be used, and there is never any harm in applying. Rarely are these funds available for non-presenters, so you probably will need to have a paper accepted to win any funding.

16.2.4 Presenting

The real reason, of course, why you want to attend a conference and present is to network and showcase yourself and your work to potential employers or postgraduate degree supervisors, or future colleagues. Nobody needs you to pretend otherwise, but at the same time, you obviously want to do the best possible job of presenting your paper, so as soon as you have been accepted for a conference, start preparing your presentation (and, if required, written paper).

You should have been given presentation practice and instruction as part of your degree course(s), and most likely for a short project presentation. But, it's quite possible that this will be the longest presentation you've ever given, and potentially the most important. So, here are some hints as you put it together:

- Keep the number of slides down. A good general rule is one slide per minute for the first ten minutes, then a maximum of one per three minutes thereafter. So, a 30 minute presentation should never use more than 17 slides as an absolute maximum.
- When you're presenting, they have come to hear you talk, not read from your slides. So make your material simple, visual, and any text large and easy to read.
- Assume your audience does have good general subject knowledge, and also that they have no interest in your university's specific academic requirements. So stick to the particular things done, and learned, on your dissertation project.
- It is always valid to include a favourable mention of anybody else who contributed to the work—your supervisor and stakeholders or sponsors/client(s) particularly. Remember they may have friends and colleagues in the audience!
- If there are going to be questions, consider what they might be and have prepared after your "last" slide several that you can jump to that provide interesting material which you can refer to in answering those questions.
- Don't forget to ensure that your last slide (usually an attractive and relevant image and something like "discussion", "questions" or "thank you" on it) includes your email address, and is up long enough for people to copy it down.
- Make sure that you got permission from anybody else (most likely the project client) to include any material that they have a stake in.
- If you produced any kind of artefacts in your dissertation project, don't be afraid to take them to the conference—just don't hand to the audience anything fragile.
- The most common failing of inexperienced presenters is trying to include much too much information. Only present as much information as the time, and the topic, need—no more.

Preparation is everything for an important talk: so practice, practice, practice. If you have friends or colleagues you can practice in front of, even better (although get it quite good first, don't inflict multiple development versions on them unnecessarily). Filming and playing it back to yourself is also a good idea. It is vitally important that you develop and practice your talk that you can deliver it within time, which means usually between about two minutes early and finishing exactly on the deadline. Conferences always take a dim view of presenters who overrun—make sure it's not you.

How to present yourself will vary between fields of engineering, and whether it's a professional or academic conference. But, unless you have clear information otherwise—we're engineers, and we turn up in business attire! Certainly turning up in a suit where everybody else is wearing jeans and T-shirts will cause you far fewer problems than the other way around. Hopefully, however, you can get good advice here from colleagues or supervisors who have attended similar conferences before— and many conferences will provide some guidance in the joining instructions. One thing that can often go down well is "teamwear", where you were part of a project

Fig. 16.3 Example of a
good business card for a new
or soon-to-be engineering
graduate

Beatrice Shilling, BEng(Hons) ARAeS	
Aeronautical Engineering Graduate	Email: BeatriceS@jetengineer.com
(propulsion engineer)	Cell: +44 7777 123 4567

team who created their own branded clothing: but again, check. Whatever you turn
up wearing, make sure it's clean and in good condition.

16.2.5 Conference Follow-Up

Don't forget to take business cards to a conference—if you're a student or recent
graduate, they don't need to be elaborate (indeed, absolutely should not be elaborate
or pretentious). Something like the example in Fig. 16.3 is excellent and such cards
can be obtained cheaply enough from various online suppliers or local shops. When
you meet people who have common interests, absolutely do swap cards. It is considered normal and healthy a few days or weeks after a conference to follow-up contacts
if there were common interests: usually by email, or even by phone if the contact
was particularly friendly or that was invited. Obviously, when following up anything
but a purely social contact, have clear objectives when you do that, whether that's
discussion about the presentation you, or they, gave; information about a possible job
opening or PhD studentship, collaboration on a future project (such as a paper). Keep
contact brief, polite and play also to whatever are the interests of the person you're
contacting. Many follow-ups will go nowhere, but some will prove useful, and this
is an important skill you're likely to need throughout any professional career.

16.3 Publishing in Academic Journals and Books from Your Dissertation

16.3.1 About Publishing in Academic Journals

Writing for peer-reviewed journals is a big deal in academia, and even in industry
can have significant benefits, in terms of exposure, building your curriculum vitae
or resumé, and providing public credibility for a piece of research work. So, there
are good reasons, if you ever can, to be published in these journals—and there are
hundreds of journals in a whole range of engineering sub-fields available in which

it's possible to publish. You would have read a reasonable number of such journal papers in the course of your dissertation, so you will be familiar with their general content and format (and that both can vary quite a lot between journals.) If it's your ambition to become an engineering researcher—particularly to progress to an MPhil or PhD, then having a published paper will massively aid you in gaining a place.

However, this is also a difficult thing to do, and very often even excellent dissertation work, deserving of a solid A-grade, may be unsuitable for publication in such journals. Any reputable journal requires two things from a paper (assuming that it's also in scope for the journal's subject area):

(1) That it is the right length and format for that journal.
(2) That the work genuinely constitutes new knowledge, worthy of *archival publication*.

No student dissertation, as written, will immediately meet the first requirement. So, to turn your dissertation: likely to be 10,000–15,000 words, formatted to meet a university's requirements into a 3,000–5,000 words journal paper, formatted to meet a journal's requirements—will take a substantial amount of effort. Experienced academics whose number of papers is in the tens or hundreds expect to spend weeks on this task and, not having done it before, you'll take a lot longer.

Secondly, however proud you are of your dissertation work (and hopefully you are) in a majority of cases, it does not constitute new knowledge. Design studies—which is a large proportion of engineering dissertations—however good, just aren't definable as archival quality research. Characterization of a new structure, shape or programme in terms of flow, strength or function usually is not in itself also of archival quality—unless, just maybe, it's incorporated into a broader study of how and why that was designed, and its potential applications.

16.3.2   Involving Others in Writing a Paper

Some undergraduate and particularly MSc work in engineering does get published, almost invariably in collaboration with more experienced academics, and if you think that your work is likely to meet the "new knowledge" criteria, this is worth exploring. Almost certainly the first person you want to talk to about this is your supervisor—they are an experienced academic who have already published in academic journals—probably many times—themselves. Their contract with the university usually requires them to publish regularly, in high-quality journals—so they have a vested interest in doing so.

Sometimes also students (including PhD research students) and others (e.g. your project client, postdoctoral researchers or stakeholders who have supported your work) may be working on different parts of a bigger project and whilst no individual part is necessarily adequate for publication: a sum of parts *may* be.

So, if you think that you have work (where "you" may actually be a group) deserving of archival publication, the best thing to do is talk to your supervisor particularly, and also anybody else who has made contributions to projects. "Solo" papers are more common in engineering than in most other fields, but nonetheless are not normal nowadays and in particular, as a relative newcomer to the research game, you will benefit greatly from collaboration with more experienced players. The experienced and already published academics will have the most sound judgment on whether work has a chance of being published.

When you go to colleagues with an idea for publication, it's a good idea to go with some ideas, specifically:

- What journal? (look for papers with similar themes: as a first guide, which journals feature in the most references in your dissertation report.)
- What structure do you propose for the paper?
- Is new data or analysis needed beyond what already exists to fit that structure?
- Who is going to do the work of manipulating the available material into a paper? Have they really got the time available?
- Who should the authors be, and in what order? (As a working rule, the first author is the person who did the most work on it, the last is the most senior who was involved—often the project supervisor; others usually in order of contribution. Another, more egalitarian, approach is simply to declare everybody equally involved and list them in alphabetical order.)

Offer your thoughts, then listen hard to the most experienced academics—if they think that the work is unsuitable for publication then, sadly, they're probably correct. If they are enthusiastic and prepared to work with you on this, well done! Now move ahead and take their guidance on how to turn your work into a publishable paper.

16.3.3 Beware of Predatory Publishers

A result of the ongoing move towards open access and online publishing of academic journals has been the invention of what are termed "predatory journals". These may often appear to be respectable academic journals, but in reality have been set up purely to exploit ambitious academics, who need to meet publication targets, without actually having any proper and robust peer-review process. The existence of these was first publicized by an American university librarian, Jeffrey Beall, in "Beall's List". Mr Beall no longer publishes that list, but searching online for the terms "Beall's List" or "Predatory Journals" will easily find you, usually anonymously published, updated lists of thousands of dubious journals and descriptions of how to spot them. Invariably, they also have no published, or a negligibly small, impact factor.[1] There is negative value in publishing in most of these predatory journals, as it will cost money, prevent you later publishing the work in respectable journals

---

[1] Impact factor, or IF, matters a lot to many universities as a measure of respectability of a journal. There are several variations on this, but the most common is a mean of the number of citations in subsequently published paper, each paper achieves within three years. Typical good engineering journals usually have an IF in the range 0.4–0.9, whilst a few can achieve higher values, perhaps

Fig. 16.4 'A busy afternoon in a publisher's office' (by Mr William Heath Robinson)

and potentially have you labelled by your peers as a gullible fool. So, do your own diligence, before submitting to any journal and avoid at all costs submitting to these predatory publishers.

Similarly, students who have published their dissertation report, particularly if you did so online, either on a personal webpage or a university research archive may well receive invitations to publish your work as a book. Almost certainly when you check the details of this "offer" it requires you to pay for this service. Very rarely does this offer you any value at all. Putting your work online is a great idea, and if you really think that you've done something of value as a book reformat it yourself and either offer it to a publisher (Fig. 16.4) or self-publish using a "publication on demand" service such as Lulu, Blurb or Kindle Direct. As a rule, anybody who is prepared to publish your dissertation, unmodified and charge you for doing so is just after your money and will add absolutely no value to your life and career.

16.3.4   In Summary

For some students, there may be possibilities and good reason to publish *from* your dissertation—particularly in academic journals. To do so, almost certainly means

---

as high as 5.0. The highest ranked scientific journals can approach 20—reflecting that published papers are far more the stock in trade of science, than they are for engineering.

significant additional effort, and collaboration with others: likely to include your supervisor and project client.

However, a lot of high-quality student work won't be suitable for onward publication—and if it's not, don't get despondent, just move on.

There are predatory publishers—both of books and journals who will try to take your money to publish material in forms that nobody will ever buy, read or cite—and you should take care not to get involved with them.

If you do manage to get your work published in a good peer-reviewed academic journal, then this can be highly beneficial—particularly if you plan your future career to remain in academia.

Appendix A
Referencing, Citing and Quoting

A.1 Introduction

It is inevitable and essential that any engineering report will contain material from and references to other earlier documents. This can be the results of mathematical proofs, borrowed diagrams, quotations or simply references to where information was found. It is really important that when doing so, this happens in an appropriate and consistent way. Failure to do so at best creates a clumsy hard to read document, and at worst makes it look either padded or plagiarized. We all know, nowadays, about the fact that nobody approves of plagiarism, and strong evidence of it in something as important as a dissertation report could end in being expelled from the university—so it's vital not to even raise suspicions of it.

Please note that we've already discussed this topic, to some extent, in Chap. 5— the opening literature review. This appendix repeats some of that information, but then expands it to provide a single point of reference.

A.2 Sources of Information

Information that you might be using can come from websites, books, conference presentations, conference papers, interviews, learning software, lecture notes, and museum placards, the list of sources is long, complex and getting more so all the time. Often, it's only necessary to show where information comes from, or sometimes to actually include some of it in the report. Either way, the following rules are always essential. You must:

- Show where the information came from
- Describe it clearly and consistently with all other citations in the document
- Include only enough information to serve the reader, and no more.
- Provide the enabled reader to find the source document, in case they want to read it more fully.

© Springer Nature Switzerland AG 2020
P. Gratton and G. Gratton, *Achieving Success with the Engineering Dissertation*,
https://doi.org/10.1007/978-3-030-33192-4

One thing that, nowadays, needs to be clearly stated and often repeated is that whilst you can find lots of useful information on the internet, being on the internet does not necessarily make it a website. So, if for example you want to cite an original source for the theory of evolution, an extremely famous source is Charles Darwin's "On the *Origin of Species.*" Being out of copyright this is readily found online; for example, a first edition at: https://www.gutenberg.org/ebooks/1228. Whilst that is a valid URL (or *Uniform Resource Locator*) for a website where you can read and download the first edition of Darwin's famous book, the book is still a book. So a much more correct citation would be:

> C Darwin, On the Origin of Species: or the preservation of favoured races in the struggle for life, John Murray (London) 1859.

There are two fundamental ways to understand the difference. First, if a source online might be routinely updated and retain the same URL, then it is probably a website (e.g. an online news article on the BBC, or an online engineering database). Secondly, if there is an available way of citing a document which does not rely upon a URL, then it probably should not be cited as a website.

An alternative to URLs, often used for online academic sources are DOI, or digital object identifiers. In theory, if not always in practice, these are non-expiring and only apply to a document which will not be changed—therefore appending a DOI to a reference for a paper may often (unlike using URLs) be a good practice.

Unless a specific style guide says otherwise, report e-books in the same way as normal paper books, except that if they exist at a unique URL, then that should also be stated after the main reference information, along with (if there is any reasonable possibility that the e-book may be modified with time) the access date.

If a book has been translated from another language, it's normal after the title to put ".trans A.Translator" *(where A.Translator is the name of the translator)* after the title, unless a style guide recommends a different way of saying this.

If a document originates with an organization rather than a named individual, use the organization as the author's name, but don't re-arrange any words. So NASA is still the "National Aeronautics and Space Administration" not "Administration, N.A.S" and the New York Times is still the "New York Times" not "Times, NY". Whilst in a journal, it's normal for reasons of space to abbreviate organization or journal names, it's seldom necessary in a report, including a dissertation report—so unless told otherwise, it's usually best for example to leave "Aeronautical Journal" as it is, not shortened as you may often see it to "AeroJ", nor the New York Times to "NYT".

Above all else, whatever system you use, be consistent with yourself within the same document. If you ensure that, and provide enough information, most other problems will be forgiven by your reader.

A.3   The Most Common Citation and Referencing Methods

Below, we cite the same list of documents in each style so that you can see and understand the differences. The list is not exhaustive—many tens, probably hundreds, of referencing systems and sub-systems exist. More detailed *style guides* for most systems are widely available, and it's a good idea if you know what system you should be using to have one of those to hand, as well as this appendix. Also some citation tools may automate this for you (although not always very well), as well as some referencing websites (bibsonomy.org being one such, worldcat.org another). If you want to cite in, for example, a book or report, only a specific section, page range or chapter, it is normal and acceptable to just add for example "12–36", or "Chap. 4" after the reference.

It is likely that your university will state a particular method that they prefer you to use, but some will not. What is paramount in all cases, however, is that you use the same citation format throughout any single document or project. However, if you'd like our suggestion—if nothing else is imposed upon you, use the AIAA system (there is also, not detailed here, a very similar IEEE or Institution of Electrical and Electronic Engineers system you may come across).

You might wonder what the difference is between a citation, and a reference? The citation is the component within the text, and points to the reference, which is at the end of a chapter or document.

Some style guides may often propose very minimalist views of referencing—for example, for a broadcast they may propose only listing the name of a radio broadcast, but not who was interviewed, and not (if it is available online) a URL. We don't agree—the whole point of referencing is to allow the reader to both verify the information used, and expand their own knowledge if they wish: if the reader is not given good opportunity to find that information themselves, it is pointless. So, we would recommend erring on the side of too much information, if at all unsure, and ultimately always include enough information that the reader can find the reference themselves without difficulty. Hence, in some of our advice below, we have included more information than some style guides. Students who are unsure whose advice to take should talk to their supervisor, or refer to the university's own style guide: between them, those should provide definitive advice.

The *details* here are not exhaustive—there are other forms of information which might be cited and referenced, for example: software, interviews, movies, emails, magazine articles, conversations, television programmes, archives or paper encyclopaedias. Often individual style guides may be available to help with these, but if not, you're unlikely to go astray by applying the same basic principles as any other reference within the system and always remembering the basic rule that there must be enough information for the reader to be able to trace the material themselves.

AIAA

The American Institute of Aeronautics and Astronautics has a wide range of highly influential journals, and this influence has extended to its system of referencing being widely used also. It is well suited to engineering journals or reports, and most journals in fields related to electronic, mechanical or aeronautical engineering use something very similar. Call-outs or citations from the text are done numerically in square brackets (e.g. [27]), and the references are listed in the order they appeared in the text (rather than in alphabetical order as used, for example, by Chicago and Harvard) which tends to lend itself to easy document automation using modern word processors. Where a DOI exists for a paper, it should always be given, as should page numbers if they are available and relevant. Normally author lists aren't abbreviated.

The order of an AIAA reference is:-

- Authors (surnames then initials)
- "Title"
- *Name of publication or conference in italics*
- Name of publisher or originating organization
- Except for whole first edition books, edition numbers of books.
- Any discrete document reference information
- (Publication date)
- [Access date] if applicable.

Book: [1] Darwin, C., "On the Origin of Species", John Murray (1859.)

Book chapter: [2] Darwin, C., *On the Origin of Species*, John Murray, 6th edn 1872 Chap. IV

Report : [3] Presidential Commission on the Space Shuttle Challenger Accident. "Report to the President". Washington, D.C. (1986)

Journal paper: [4] Blackawton, P. S., Airzee, S., Allen, A., Baker, S., Berrow, A., Blair, C., Churchill, M., Coles, J., Cumming, R. F.-J., Fraquelli, L., Hackford, C., Hinton Mellor, A., Hutchcroft, M., Ireland, B., Jewsbury, D., Littlejohns, A., Littlejohns, G. M., Lotto, M., McKeown, J., O'Toole, A., Richards, H., Robbins-Davey, L., Roblyn, S., Rodwell-Lynn, H., Schenck, D., Springer, J., Wishy, A., Rodwell-Lynn, T., Strudwick, D., and Lotto, R. B., "Blackawton bees," *Biology Letters*, vol. 7, (2011), pp. 168–172. https://doi.org/10.1098/rsbl.2010.1056

Newspaper: [5] New York Times, "Men walk on Moon", *New York Times*, Vol CXVIII no. 40, 721, pp 1 (July 21 1969)

Webpage: [6] Civil Aviation Authority, "Unmanned Aircraft and Drones". Available at https://www.caa.co.uk/Consumers/Unmanned-aircraft-and-drones/ [Accessed Jan. 01 2020].

Online Reference: [7] Encyclopaedia Britannica Online, "Isambard Kingdom Brunel". https://www.britannica.com/biography/Isambard-Kingdom-Brunel [accessed July 11 2019]

Broadcast: [8] *Desert Island Discs*, "Professor Dame Anne Dowling", BBC, (Broadcast 29 July 2016)

Lecture Notes: [9] Gratton, P.M. (2016), "Management and Financial Aspects-payback on NZEB retrofit", *Lecture on course ME5123*, Brunel University London (2016)

APA

APA (or American Psychological Association) method was first created for use in science in 1929, and nowadays, it is most commonly used in the social sciences; it is a complete writing system presently in its 6th edition, but here we only discuss citing and referencing. APA versions have different approaches to abbreviation of authors' lists: most common is to include all authors, but some versions will only include the first six, which we have done with our journal example here, and some only the first author. Calling out from the text is normal using surname and text; for example (Darwin 1859) as with Chicago, et.al.

The normal order of an APA reference is:-

– Author (surname, first initial); for multiple authors normally only the first six followed by "et al."
– Year of publication
– *Title* (*in italics* for long works such as a book, "in parentheses" for shorter works such as a journal paper).
– Location of publication
– Publisher
– (If applicable) URL or DOI

Book: Darwin, C . (1859). *On the Origin of Species*. John Murray.

Journal paper: Blackawton, P. S., Airzee, S., Allen, A., Baker, S., Berrow, A., Blair, C., et.al. (2011). "Blackawton bees". *Biology Letters*, 7, 168–172. https://doi.org/10.1098/rsbl.2010.1056

Report : United States. (1986). *Report to the President*. Washington, D.C: Presidential Commission on the Space Shuttle Challenger Accident.

Newspaper: New York Times (1969) "Men walk on Moon", New York Times, Vol CXVIII no. 40, 721, p. 1

Webpage: Civil Aviation Authority (2015), "Unmanned Aircraft and Drones". Available at [Retreived Jan. 01 2020 from https://www.caa.co.uk/Consumers/Unmanned-aircraft-and-drones/].

Online Reference: Encyclopaedia Britannica Online (2019), "Isambard Kingdom Brunel", accessed July 11 2019, https://www.britannica.com/biography/Isambard-Kingdom-Brunel

Broadcast: Desert Island Discs (2016), "Professor Dame Anne Dowling", BBC, Broadcast 29 July 2016

Lecture Notes: Gratton, P. (2016), "Management and Financial Aspects-payback on NZEB retrofit", Lecture on course ME5123, Brunel University London Brunel University London

Chicago

The Chicago system is widely used in humanities, so not strictly normal in engineering—but it can be adapted for engineering writing, and some universities impose it on engineering and science departments in the name of standardization. It is very well described at www.chicagomanualofstyle.org.

Chicago referencing, so far as possible, includes:-

– Author's names as normally expressed
– Title
– Publication name
– Publication year
– Publication month and date
– Publisher
– City of publication
– Date of access (if it is an online source)
– Page numbers (if not the whole document)
– URL or DOI (for some online sources)

The list of references at the end of the document should be listed by alphabetical order of author's surnames, single space, and the second and subsequent lines of each entry indented. Within the text, citations are written as "(Surname YYYY)". Where you use the same source repetitively this should be shown as "Ibid", which is an abbreviation of the Latin "Ibiden" or "In the same place".

Book: Charles Darwin, *On the Origin of Species: or the preservation of favoured races in the struggle for life* (London, John Murray, 1859)

Report : United States. 1986. *Report to the President*. Washington, D.C.: Presidential Commission on the Space Shuttle Challenger Accident.

Journal paper: BlackawtonPrimary School, Airzee, S., Allen, A., Baker, S., Berrow, A., Blair, C., Churchill, M., Coles, J., Cumming, R. F.-J., Fraquelli, L., Hackford, C., Hinton Mellor, A., Hutchcroft, M., Ireland, B., Jewsbury, D., Littlejohns, A., Littlejohns,

G. M., Lotto, M., McKeown, J., O'Toole, A., Richards, H., Robbins-Davey, L., Roblyn, S., Rodwell-Lynn, H., Schenck, D., Springer, J., Wishy, A., Rodwell-Lynn, T., Strudwick, D. and Lotto, R. B.. "Blackawton bees." *Biology Letters* 7, no. 2 (2011): 168-172.

Newspaper: Men walk on moon, New York Times, 1969 July 21, p.1

Webpage: Civil Aviation Authority, Unmanned Aircraft and Drones, accessed January 01 2020, https://www.caa.co.uk/Consumers/Unmanned-aircraft-and-drones/

Online Reference: Encyclopaedia Britannica Online, "Isambard Kingdom Brunel", accessed July 11 2019, https://www.britannica.com/biography/Isambard-Kingdom-Brunel

Broadcast: BBC Desert Island Discs, "Professor Dame Anne Dowling", Presented by Kirsty Young, Produced by Sarah Taylor, Broadcast 29 July 2016, https://www.bbc.co.uk/programmes/b07lf8fw

Lecture Notes: Petra Gratton, "Management and Financial Aspects-payback on NZEB retrofit", Lecture on course ME5123, Brunel University London February 01 2016

Harvard

The Harvard referencing system is probably the most common used in non-humanities academic work, so is most likely to be the basis of style guides used in university engineering and science departments. Harvard is similar to Chicago in general principles (although not in detail) and follows the order convention:-

– Author(s) names(s) given as family name then initials
– Publication year *(in brackets)*
– Title
– Publication location (if relevant)
– Publisher
– Except for whole first edition books, edition numbers of books.
– Specific location (e.g. page numbers)

Like Chicago, Harvard lists references in alphabetical order of the authors names, and cites them by name and publication year, (e.g. Darwin 1859)

Book: Darwin, C. (1859), *On the Origin of Species*, John Murray
[In Harvard , we don't list the edition for first editions if the whole book is referenced, but do for subsequent editions, or where only part of the book is cited, so....]

Book chapter: Darwin, C. (1872), *On the Origin of Species*, John Murray, 6th edn Ch.VI.

Report :   UNITED STATES. (1986). *Report to the President*. Washington, D.C., Presidential Commission on the Space Shuttle Challenger Accident.

Journal paper: Blackawton, P. S., Airzee, S., Allen, A., Baker, S., Berrow, A., Blair, C., Churchill, M., Coles, J., Cumming, R. F.-J., Fraquelli, L., Hackford, C., Hinton Mellor, A., Hutchcroft, M., Ireland, B., Jewsbury, D., Littlejohns, A., Littlejohns, G. M., Lotto, M., McKeown, J., O'Toole, A., Richards, H., Robbins-Davey, L., Roblyn, S., Rodwell-Lynn, H., Schenck, D., Springer, J., Wishy, A., Rodwell-Lynn, T., Strudwick, D. & Lotto, R. B. (2011). Blackawton bees. *Biology Letters*, 7, 168–172. https://doi.org/10.1098/rsbl.2010.1056

Newspaper: New York Times (1969) *Men walk on Moon*, New York Times, Vol CXVIII no.40,721, pp1

Webpage: Civil Aviation Authority (2015), *Unmanned Aircraft and Drones* [online]. Available at https://www.caa.co.uk/Consumers/Unmanned-aircraft-and-drones/ [Accessed Jan. 01 2020].

Online Reference: Encyclopaedia Britannica Online (2019), *Isambard Kingdom Brunel,* accessed July 11 2019, https://www.britannica.com/biography/Isambard-Kingdom-Brunel

Broadcast: Desert Island Discs (2016), [Radio programme] *Professor Dame Anne Dowling*, BBC, Broadcast 29 July 2016

Lecture Notes: Gratton, P.M. (2016), "Management and Financial Aspects-payback on NZEB retrofit", Lecture on course ME5123, Brunel University London

ISO 690

ISO 690 is an international standard for document citations, replacing an older DIN 1505-2. It has not been much taken up in the English speaking world, but is more common particularly in German and some other European languages. We have shown the "Author-date" English language form of ISO690 here—if you are writing in another language, it would be wise to look up the variation used in that language, of which there are several for examples, French, German, Czech, and so on (the system originated in a German language system). Similar to AIAA or Vancouver, ISO 690 uses numeric citations. Some versions capitalize author names, but we haven't done so here. Unusually, ISO690 doesn't require access dates for web resources.

The usual order of an ISO690 reference is:

– Author's surname then initial(s)

– Date of publication
– Title
– Name of publication (in italics)
– Discrete publication reference information
– If applicable, a URL or DOI.

Book:	1. Darwin, C., (1859). On the origin of species. John Murray.
Journal Paper:	2. Blackawton, P. S., Airzee, S., Allen, A., Baker, S., Berrow, A., Blair, C., Churchill, M., Coles, J., Cumming, R. F.-J., Fraquelli, L., Hackford, C., Hinton Mellor, A., Hutchcroft, M., Ireland, B., Jewsbury, D., Littlejohns, A., Littlejohns, G. M., Lotto, M., Mckeown, J., O'toole, A., Richards, H., Robbins-Davey, L., Roblyn, S., Rodwell-Lynn, H., Schenck, D., Springer, J., Wishy, A., Rodwell-Lynn, T., Strudwick, D. and Lotto, R. B., 2011. Blackawton bees. Biology Letters (2011). Vol. 7, no. 2, p. 168–172. DOI https://doi.org/10.1098/rsbl.2010.1056. Available from: http://rsbl.royalsocietypublishing.org/content/7/2/168.abstract
Report :	3. Presidential Commission on the Space Shuttle Challenger Accident. (1986). Report to the President. Washington, D.C.
Newspaper:	4. New York Times (1969), Men walk on Moon, New York Times, Vol CXVIII no.40,721, pp1
Webpage:	5. caa.co.uk., Unmanned Aircraft and Drones. 2015. Retreived from https://www.caa.co.uk/Consumers/Unmanned-aircraft-and-drones/
Online Reference:	6. Encyclopaedia Britannica Online, "Isambard Kingdom Brunel",, https://www.britannica.com/biography/Isambard-Kingdom-Brunel
Broadcast:	7. Desert Island Discs, Professor Dame Anne Dowling, BBC, (2016, 29 July)
Lecture Notes:	8. Gratton, P.M., Management and Financial Aspects-payback on NZEB retrofit, Lecture on course ME5123, Brunel University London (2016, 1 February)

Vancouver

The Vancouver system originated in biological and medical sciences, but is very similar to the AIAA system, albeit that references are called out by citations given as subscripted small numbers. Where a URL or DOI for a document is available, it should be provided after the main reference wording. So, the order of a Vancouverreference is:-

– Authors (surnames then initials), long lists usually being truncated early

– *Title in italics*
– Except for whole first edition books, edition numbers of books.
– Name of publication or conference
– Name of publisher or originating organization
– Any discrete document reference
– Publication date
– Access date, if applicable, given as DD/MM/YYYY.

Book: (1) Darwin, C. *On the Origin of Species: or the preservation of favoured races in the struggle for life.* John Murray; 1859.

Book chapter: (2) Darwin, C., *On the Origin of Species: or the preservation of favoured races in the struggle for life.* 6th ed, John Murray, 1872 Chap. IV

Report :   (3) Presidential Commission on the Space Shuttle Challenger Accident.. *Report to the President.* 1986

Journal Paper: (4) Blackawton PS, Airzee S, Allen A, Baker S, Berrow A, Blair C, et al. *Blackawton bees.* Biology Letters [Internet]. 2011;7(2):168–72. Available from: http://rsbl. royalsocietypublishing.org/content/7/2/168.abstract

Newspaper: (5) New York Times, *Men walk on Moon*, New York Times, Vol CXVIII no.40,721, pp1 (July 21 1969)

Webpage: (6) Civil Aviation Authority, *Unmanned Aircraft and Drones* . Available at https://www.caa.co.uk/Consumers/Unmanned-aircraft-and-drones/ [Accessed 01/01/2020].

Online Reference: (7) Encyclopaedia Britannica Online, *Isambard Kingdom Brunel*, https://www.britannica.com/biography/Isambard-Kingdom-Brunel [Accessed 11/07/2019]

Broadcast: (8) Desert Island Discs, *Professor Dame Anne Dowling*, 29 July 2016, British Broadcasting Corporation

Lecture Notes: (9) Gratton, P. *Management and Financial Aspects-payback on NZEB retrofit*, Lecture on course ME5123, 2016, Brunel University London

A.4   Quoting from Sources

Thereare many good reasons to quote from a source; this is most likely to be in the form of words, equations or figures. When doing so, it's essential to do two things: first always fully and correctly state the source; second, never reproduce any more

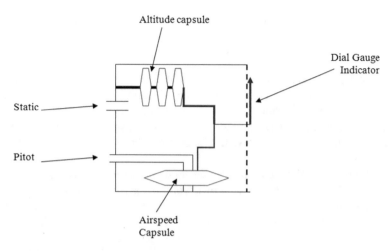

Fig. A.1 Workings of a Mach Number indicator (from Ref. [1])

material from elsewhere than absolutely necessary. Here are three examples (and below we have shown a short reference section in AIAA format):

Example 1—A diagram (see Fig. A.1)

Example 2—A quotation
K.R.G. Browne says (in reference [2]) of William Heath Robinson: "Almost anybody, given a little time for reflection, can invent a bag or a boot; even a cab is not really hard to evolve if one is reasonably hackney-minded; but only Heath Robinson could do what Heath Robinson does. This illustrates the unique characteristics of Heath Robinson's artwork, in illustrating fanciful engineering concepts."

Example 3—An equation
Reference [3] shows that $\frac{d(\coth u)}{dx} = -csch^2 u \frac{du}{dx}$.

References

[1] Gratton, G, "Initial Airworthiness: Determining the Acceptability of New Airborne Systems" 2nd edition, Springer (2018) p. 66
[2] William Heath Robinson, "Devices", Hutchinson (London) (1977) p. 6–8.
[3] Cault J & Ayres F, "Calculus" 2nd (SI / metric) edition, McGraw Hill (1972) ch. 15 p. 75

A.5 Final Comments on References and Quotations in a Dissertation

A good number of references are essential in a dissertation—typically in the order of 30–60; this should contain a combination of journal papers, teaching material, reports, standards and textbooks containing basic theory. Remember that you are not just writing an engineering report, you are demonstrating that you have fully absorbed and understood the relevant subject matter, and are capable of using a wide range of sources.

A further issue is that you must demonstrate that the report, and the dissertation project behind it, is substantially your own work. Some students have been known to try and pad their dissertations with excessive amounts of material from references, and markers are alert to that. Certainly, other people's prior work will add nothing to your marks—even figures. So it is best, whilst using a good range of references, to keep to a low minimum the amount of quoted material, and wherever possible also re-draw diagrams where this is reasonably practicable, as it also demonstrates a useful engineering skill. Even mathematical proofs, however relevant, don't really need to be quoted—stick to the final result, then progress by showing your own original thinking. As a working rule, if more than 3% of the total content of the report is not original, it becomes suspicious, and 6% or above will be very problematic. And of course, ensure that absolutely every piece of copied or quoted material is always accompanied by a clear citation.

With regard to the references themselves, remember that there is a difference between a reference list (which is all called out from the body text in citations) and a bibliography (which is simply a list of relevant literature). Most engineering reports only want a reference list, and a bibliography is normally only needed if specifically asked for by university guidelines.

Appendix B
Using Engineering Standards

B.1 Introduction

Engineering standards are an underlying fact of their working life for most professional engineers. With multiple sub-disciplines, and multiple national and international standards bodies, we can't provide in this book much insight into any individual standards: that is very much down to your lecturers and professors. However, the dissertation can sometimes be a student's first significant contact with the use of standards, and so this appendix aims to provide an insight into how standards are used in engineering, how they might be used in some kinds of dissertation projects, and what students should take from this into their subsequent careers.

Whilst not all dissertations will use engineering standards, many will. This is valuable experience for students as not only should this provide them with opportunities to demonstrate professional awareness to supervisors and markers, and thus raise their grades, but also provides a valuable learning opportunity. The graduate who understands their industry sector's main standards and how they're used is a better educated and employable graduate.

B.2 The Basic Types of Standard

Most standards fall into two categories—designstandards, and regulatory standards. They may often overlap, but we'll describe them here separately.

Design standards are those created usually by national standards bodies, such as ASTM (formerly known as American Society for Testing and Materials) or BSI (the British Standards Institute). The most useful will be merged with other national standards, to create an *international standard* For example, the international standard. for manufacturing quality assurance, ISO 9001 was developed by the International Standards Organisation (ISO) from a British Standard: BS 5750 part 1. When this happens, the original national standards are usually abolished worldwide so that only

© Springer Nature Switzerland AG 2020
P. Gratton and G. Gratton, *Achieving Success with the Engineering Dissertation*,
https://doi.org/10.1007/978-3-030-33192-4

the single international standard remains. Mostly the process of creating them is one of *consensus standards*—where committees of experts from the relevant industry sectors meet to write and approve what is essentially a statement of best practice. Unsurprisingly, when this process happens in different countries, competent engineers will usually come up with very similar solutions wherever they are—making it quite easy to eventually converge them. Compliance with designstandards is usually monitored within the industry, and it is not a mandatory requirement.

Regulatory standards are usually set in law or regulation, and are mandatory for many activities. Generally, they encompass activities where failure threatens public health and safety, for example, food packaging regulations, aviation safety or building fire protection. Engineers or their employers who breach these may be at risk of prosecution, and even imprisonment, so understanding them is important.

It is not unusual for what commence as designstandards to become a regulation when governments decide, normally in the interests of public safety, to mandate them. At this point whilst the regulations don't usually change, the means of demonstration will—engineers may start to have to lodge proof with independent bodies, or be subject to external demonstration and audit. It's also not unusual that companies, design organizations, and so on, have their own designstandards, which ensure output standardization is treated internally as if they were regulatory standards.

Usually the need for engineering products to demonstrate compliance with standards is most obvious at the design and manufacturing stages—although products subject to regular use and maintenance (e.g. a power station, or an aeroplane) may be also subjected to regular re-evaluation.

B.3   Knowing Your Standards

Does an engineering student need to know theirengineering standards? This depends upon what they're working on. If, for example, your project involves developing or evaluating the design of all or parts of a car, an aeroplane or a building, then you should have found and understood the relevant standards as part of the literature review. Alternatively, a pure researchproject into, for example, an aspect of fluid mechanics *may* not have had much use for standards, although standards may still exist for instrumentation, report writing and so on. So, the first assumption should be that there are relevant standards for any engineering activity, including a university dissertation, unless it's been conclusively shown that there aren't.

All engineers working on all projects must understand the applicable standards to their work, and do so as early as possible. At best, modifying something already done to meet standards after the majority of design and construction work has been done will be expensive and time-consuming; at worst, it may be impossible with the resources available. Some engineering disciplines may consist entirely of product assessment against formal standards: for example, flight test engineers, computer validation engineers, or building control surveyors.

If your university has good library facilities, you should have access to the relevant standards: whilst some are free and in the public domain, most standards institutions charge a fee to access material. This is necessary, to them, as it's the way their work is funded. Your university will be paying a subscription to them which provides students and staff with access for as long as they are members of the university. Note that projectclients, if they are from outside the university, and particularly if they don't work within a large organization, may not have this access. All should take care that any sharing (similar to downloaded journal papers from non-open-access sources) doesn't inadvertently breach any licensing arrangements. This is also a reminder to students that once you graduate, you may not have access to these standards any more unless they've been paid for by your graduateemployer. This doesn't mean, however, that you might not be expected to know your way around the industry sector's key standards both in jobinterviews, and after starting.

B.4 Compliance Checklists

When using standards, at either the design or completion stage, you may often be asked to produce a *compliance checklist*. This is not really a checklist; it is a document where the author reviews a design against the design or regulatory standard, point by point, and provides the evidence and a concluding statement for each point demonstrating that either:

(1) The design conforms to the standard, or
(2) The design does not conform to the standard—and what will be done about it, or
(3) The design has not yet been shown to confirm to the standard—and what will be done to demonstrate conformity, or exceptionally:
(4) That the standard is inappropriate to the specific design, but proposing an alternative (and equivalent or higher) standard and how this will be met.

Compliance checklists are really useful through any design or design-and-make project, because by creating one at the first stage, then periodically reviewing it, the design team can ensure both a standard compliant design, and sufficient evidence to prove that compliance. This is part of how many designteamswork in many industries; so, it is also a prime example of how the dissertation helps students learn skills and develop knowledge that will be valuable in their professional career—and can be described in a jobinterview.

B.5   A Caution About Standards Overkill

In industry, there may be whole compliance departments preparing extremely long and complex demonstrations of standard compliance, and large authority departments overseeing their work. That level of complexity is unrealistic for a dissertation. The student(s) and supervisor and if appropriate the client as a subject matter expert should discuss any plans to prepare a compliancechecklist, and which standards will be used and to what depth. It's wise to agree which (of potentially dozens of) standards should be used, and the complexity of compliance demonstrations. For dissertation purposes, it's likely that just using a subset will be appropriate to the task.

Appendix C
Competitions

C.1 Introduction

There are a lot of competitions available for engineering project students that offer opportunities to engage with people from industry and other universities, to give projects a target, and to win prizes. Whilst they may not suit all projects (or students), it's worth being aware of them.

Project competitions are usually organized by bodies who are trying to achieve something, and it may not always be what it *appears* to be about. For example, for many of the students entering into Formula SAE (Formula Student in the UK)—designing and testing single-seat race car–the competition appears to be all about motorsports achievement. In reality, Formula SAE is actually about showing achievement over the whole breadth of engineering product development, from conceptual design, through detaildesign, ergonomics, performance and structural analysis, standards compliance, to prototyping, product testing and marketing. This doesn't mean that most students entering Formula SAEcompetitions aren't mostly happy motorsports enthusiasts—but it does mean that the best prizes usually go to the studentteams who understand and embrace the whole breadth of subject coverage.

The bodies sponsoring competitions may do so for multiple reasons. They may be trying to enhance engineering education, raise the profile of their industry or activity or give themselves a chance to snap up the best graduates in particular fields. All of these, of course, are good things, and well worth engaging with—but the most benefit will usually be obtained from understanding and embracing these real reasons, and not just the most obvious and glamorous aspects.

Listed below are a range of competitions that repeat regularly so should be available for students using this book to enter. If this is your plan, early discussion between the student(s) and their supervisor(s) is a very good idea to make sure that the projectobjectives and timescales fit with both the competition and the university's objectives. It is likely that you'll be treating the competition rules and organizer as

© Springer Nature Switzerland AG 2020
P. Gratton and G. Gratton, *Achieving Success with the Engineering Dissertation*,
https://doi.org/10.1007/978-3-030-33192-4

the client, although with some modification. And don't forget, winning a big competition is great—but getting great grades in your degree is greater, and learning skills that will take your career where you want it to go, greatest of all.

The financial position of competitions will vary significantly. Some are expensive to enter, likely requiring university or commercial sponsorship, whilst others are effectively free. Similarly, some competitions come with large cash prizes, whilst others come simply with the kudos of winning. Check the details of each competition to find out, and if a lot of money is potentially involved, talk to your university supervisors—they may sometimes have the ability to help teams, particularly if they will be raising the profile of the university by competing. Sometimes raising sponsorship is seen as part of the competition and associated learning experience.

It is important to recognize that, if targeting a dissertation project towards a competition, these prizes are hard to win, and there are never easy wins. Any student, or team of students, aiming at these competitions needs a genuine passion for the subject that will sustain them through long hours of additional work that will be needed to shine against competition from other universities, and often other countries. This, of course, is exactly the sort of passion and dedication that future employers will be looking for in a graduate, and why engagement with competitions is so useful as an extension of the dissertation. Consider from the perspective of an employer, the difference between the graduate who only did the minimum hours to meet their university objectives and the graduate who had developed such a passion for their topic that they worked long hours trying to achieve world beating outcomes in it. It is obvious which of these two graduates an employer is most likely to want to hire.

C.2 Project Competitions

What's the competition?	AIAA Student Design Competitions
Who organizes it?	American Institute of Aeronautics and Astronautics
When and where?	Multiple competitions, mostly US centred
Who can enter?	Any students who are AIAA members
What's it all about?	Various design challenges, including aircraft, powerplant and space systems
What are the prizes?	Typically up to $1000
Where's the website?	https://www.aiaa.org/home/get-involved/students-educators/Design-Competitions

What's the competition?	ASAIO & ASAIOfyi Annual studentdesigncompetition
Who organizes it?	American Society for Artificial Internal Organs
When and where?	Preliminary Proposals February Presentations by selected applicants June
Who can enter?	Current undergraduate students and is targeted to senior year engineering students performing their capstonedesignproject on a medical technology
What's it all about?	The design of devices, tissue engineering and cell therapy aimed at the treatment of a disease
What are the prizes?	Not known (NK)
Where's the website?	https://asaio.com/annual-conference/student-design-competition-2/

What's the competition?	ASMEStudentdesigncompetition
Who organizes it?	American Society for Mechanical Engineers
When and where?	Asia-Pacific competition: March Western USA: March Eastern USA: April
Who can enter?	Teams of ASME student members
What's it all about?	Each team is required to design, construct and operate a prototypemeeting the requirements of an annually determined problem statement
What are the prizes?	NK
Where's the website?	http://www.asme.org/events/competitions/student-design-competition

What's the competition?	American Solar Challenge
Who organizes it?	Sunrayce, US Department of Energy, National Renewable Energy Laboratory
When and where?	Across the USA on public roads (typically a 2,000 miles/3,000 km route) even numbered years about July/August. Formula Sun Grand Prix is a required qualifying event
Who can enter?	University teams
What's it all about?	Designing, building, then racing solar-powered cars across America.
What are the prizes?	NK
Where's the website?	http://americansolarchallenge.org

What's the competition?	Autocar next generation award
Who organizes it?	Autocar magazine, Courland Automotive, Society of Motor Manufacturers and Traders
When and where?	Annual, UK
Who can enter?	Students aged 17–25 either living in the UK, or studying in the UK

(continued)

(continued)

What's it all about?	Students enter summaries of their ideas and designs that are innovative in automotive technology
What are the prizes?	£8,000 and 6 months work experience
Where's the website?	https://www.autocar.co.uk/nextgenerationaward

What's the competition?	Bridgestone World Solar Challenge
Who organizes it?	State of South Australia
When and where?	October competition, odd numbered years, South Australia
Who can enter?	Anybody
What's it all about?	To promote the development of solar-powered vehicle technology, teams compete to enter wholly solar-powered vehicles racing on a 3,022 km (1,878 m) route from Darwin to Adelaide, Australia
What are the prizes?	NK
Where's the website?	https://www.worldsolarchallenge.org

What's the competition?	British Model Flying Association Payload Challenge
Who organizes it?	British Model Flying Association, CargoLogic Air, Royal Aeronautical Society, BAE Systems
When and where?	Fly-off June
Who can enter?	Teams of five university or school students (Different categories are open to different age/educational brackets)
What's it all about?	Teams must design, build and fly a model aircraft to deliver a payload carrying mission
What are the prizes?	Prizes in various categories including egg lift, distance, weight
Where's the website?	http://payloadchallenge.bmfa.org/

What's the competition?	Cleantech challenge
Who organizes it?	London Business School and University College London
When and where?	Stage 1 idea submission January Stage 2 business plan submission February Final two-day "bootcamp" in London April
Who can enter?	Anybody registered as university student anywhere in the world
What's it all about?	A competition to develop clean technology ideas with associated business plans
What are the prizes?	£10,000
Where's the website?	https://www.cleantechchallenge.uk/

What's the competition?	Design by Biomedical Undergraduate Teams (DEBUT) Challenge
Who organizes it?	National Institute of Biomedical Imaging and Bioengineering (NIBIB) and VentureWell
When and where?	Submission deadline May, Winners announced August, Award ceremony October in USA
Who can enter?	Teams of at least three students, who must be US citizens (foreign citizens may be team members but won't receive a share of cash prizes), at least one of whom must be studying biomedical engineering
What's it all about?	Teams enter designs for life-enhancing biomedical devices or technologies
What are the prizes?	Multiple prizes up to $20,000
Where's the website?	https://www.nibib.nih.gov/training-careers/undergraduate-graduate/design-biomedical-undergraduate-teams-debut-challenge

What's the competition?	Essity Universities Challenge HaFa challenge
Who organizes it?	Essity.com
When and where?	Annual, opening October, entries due April each year
Who can enter?	Teams of 1–5 undergraduate students at participating universities across 12 countries. Team members must all be from the same university
What's it all about?	Working to a project brief, design a solution to an engineering problem in the "Hankies and Facials" and Professional Hygiene industries
What are the prizes?	Award, awards dinner, international internships
Where's the website?	https://www.essityuniversitychallenge.com/en

What's the competition?	European International Submarine Races
Who organizes it?	Institute of Marine Engineering Science and Technology, Babcock, DE&S (UK), QinetiQ
When and where?	Registration opens in July, qualifying rounds (designreport) up to April, races in July in Gosport, UK
Who can enter?	Studentteams from universities (must include three qualified divers)
What's it all about?	Design, build and race flooded submarines piloted by a single scuba diver, who must be fully enclosed within the hull of the machine. All propulsion must be provided by the diver during the race
What are the prizes?	Trophy for overall winning team, various awards: top speed, agility, innovation and presentation
Where's the website?	https://www.subrace.eu

What's the competition?	Formula Student (Formula SAE)— Fig. C.1
Who organizes it?	Institution of Mechanical Engineers (in the UK)
When and where?	(Below is UK, check arrangements for other countries) Business case released November Funding applications early December Class 1 team entries January Detailed entry, May Race-off at Silverstone in July
Who can enter?	Teams of university students (all must be IMechE student affiliate members) supported by at least one academic
What's it all about?	Teamsdesign, and in higher classes build and test, a nominal prototype for a single-class single-seat racing car (Fig. C.1) Variations also exist for ultra endurance vehicles, snowmobiles, All Terrain Vehicles, and others
What are the prizes?	Multiple awards and cash prizes in different categories
Where's the website?	UK: https://www.imeche.org/events/formula-student USA: https://www.sae.org/attend/student-events Australasia: http://www.saea.com.au/ Canada: https://www.sae.org/attend/student-events/formula-sae-north The event also exists in many other countries, with their own, usually local language, websites

What's the competition?	Formula Sun Grand Prix
Who organizes it?	Sunrayce, US Department of Energy, National Renewable Energy Laboratory
When and where?	Annually in the USA
Who can enter?	University teams
What's it all about?	Designing, building, then racing solar-powered cars on a three-day closed track event
What are the prizes?	Not known
Where's the website?	http://americansolarchallenge.org/about/formula-sun-grand-prix/

What's the competition?	Go green in the city
Who organizes it?	Schneider electric
When and where?	Global: competition opens each February
Who can enter?	Young innovators
What's it all about?	Develop and present innovative ideas in topics including sustainability, digital economy, smart manufacturing and cyber security
What are the prizes?	Mentoring and travel opportunities
Where's the website?	http://www.gogreeninthecity.com/

What's the competition?	Grads in Games Search for a Star
Who organizes it?	Aardvark Swift
When and where?	Annual
Who can enter?	UK-based university students and graduates
What's it all about?	Working to a brief, design tools that will contribute to computer game function, or in some cases complete computer games
What are the prizes?	Guaranteed job interviews, items of computer hardware
Where's the website?	http://gradsingames.com/game-dev-challenges/

What's the competition?	Graduate and Student Communications Competition
Who organizes it?	Institution of Civil Engineers (ICE)
When and where?	Multiple regional competitions around the UK
Who can enter?	Teams of 3–5 students or graduate engineers
What's it all about?	Presentation based on an imaginary civil engineering project which will affect a local community, made to a panel of judges in front of a public audience
What are the prizes?	Up to £250
Where's the website?	https://www.ice.org.uk/careers-and-training/graduate-civil-engineers/graduate-awards-and-competitions#emerging-engineers-

What's the competition?	Human Powered Vehicle Challenge
Who organizes it?	American Society of Mechanical Engineers (ASME)
When and where?	Asia-Pacific competition: March Western USA: March Eastern USA: April South America: July
Who can enter?	Teams of engineering and technology undergraduate students
What's it all about?	Students must design, build and race human-powered vehicles against a set of complex design criteria
What are the prizes?	Not known
Where's the website?	http://www.asme.org/events/competitions/human-powered-vehicle-challenge-(hpvc)

What's the competition?	IMechE Railway Challenge Competition
Who organizes it?	Institution of Mechanical Engineers Railway Division
When and where?	The locomotives will be tested live at the competition weekend, which takes place in June at Stapleford Miniature Railway in Leicestershire, UK
Who can enter?	Teams of students, apprentices and undergraduates, each with an industry or academic "team supervisor"

(continued)

(continued)

What's it all about?	Participants are required to design and manufacture a miniature (10¼" gauge) railway locomotive in accordance with a set of strict rules and a detailed technical specification
What are the prizes?	Prizes exist in various categories including: reliability, energy storage, traction, ride comfort, noise, maintainability, business case, etc.
Where's the website?	http://www.imeche.org/events/challenges/railway-challenge/about-railway-challenge

What's the competition?	IMechE Unmanned Air Systems Challenge
Who organizes it?	Institution of Mechanical Engineers, Qinetiq, GKN
When and where?	Applications: October / November Preliminary Design submission: January Critical Design Review: April Flight readiness review, final fly-off: June
Who can enter?	University teams
What's it all about?	Design, build and fly an unmanned aircraft under 7 kg to safely fly then drop bags of flour onto a target
What are the prizes?	Multiple prizes including Grand Champion, Innovation, Safety and Airworthiness, Media Engagement, etc.
Where's the website?	http://www.imeche.org/events/challenges/uas-challenge

What's the competition?	International Submarine Races–Figure C.2
Who organizes it?	Foundation for Underwater Research and Education (FURE) (USA)
When and where?	Biennially in June/July at the Naval Surface Warfare Centre, Carderock Division, Maryland, USA (Fig. C.2)
Who can enter?	Teams (which must include qualified divers)
What's it all about?	Human powered underwater vehicles are designed, constructed and raced by teams of students
What are the prizes?	Trophy and $1,000 for the overall winner; various awards for speed, innovation, design, operation
Where's the website?	https://internationalsubmarineraces.org/

What's the competition?	KYOCodes
Who organizes it?	Kyocera UK
When and where?	Annual, global
Who can enter?	University students
What's it all about?	Develop apps for future electronic control applications

(continued)

(continued)

What are the prizes?	1st place 3x £1000 Apple Vouchers per team 2nd place 3x £500 Apple Vouchers per team 3rd place 3x £250 Apple vouchers per team The top 3 winning universities will also get a Kyocera MFP Best entrants also get Kyocera HyPAS app developer accreditation
Where's the website?	No current webpage at the time of going to print

What's the competition?	Louis Braille Touch of Genius Prize for Innovation
Who organizes it?	(US) National Braille Technology, Gibney Family Foundation
When and where?	Applications must be received by the end of January and can come from anywhere in the world
Who can enter?	Anybody, either individuals or groups
What's it all about?	Designing machines or technologies that support blind people in tactile literacy
What are the prizes?	Up to $5,000
Where's the website?	http://www.nbp.org/ic/nbp/technology/tog/tog_prize

What's the competition?	Pan African Robotics Competition
Who organizes it?	Pan African Robotics Competition
When and where?	Annual finals in an African City
Who can enter?	School and University undergraduate students from 22 participating African countries, working in either English or French
What's it all about?	A range of design / build / test challenges set each year
What are the prizes?	Not known
Where's the website?	http://parcrobotics.org/

What's the competition?	Passivhaus Student Competition
Who organizes it?	The Passivhaus Trust, UK
When and where?	October: launch Spring: site visit (established Passivhaus in the UK) June: judging of short-listed projects July: winner announced October: award ceremony
Who can enter?	Students at participating universities
What's it all about?	Competitors are asked to identify one UK retrofit or new build project and provide detailed design to turn this into a certifiable Passivhaus building
What are the prizes?	Free ticket to the annual UK Passivhaus Conference, where the awards ceremony will take place; winning projects display at the UK Passivhaus Conference; featured on the Passivhaus Trust website and newsletter; a certificate and gift vouchers
Where's the website?	http://www.passivhaustrust.org.uk/

What's the competition?	Royal Aeronautical Society General Aviation Design Competition
Who organizes it?	Royal Aeronautical Society General Aviation Group
When and where?	A competition is normally run every 1–2 years, normally timed to coincide with the university year
Who can enter?	Anybody can enter, whether students, academics, professionals or general public
What's it all about?	To promote innovation in light aviation, the RAeS GA group has for many years run regular competitions for aspects of light aircraft design. This has helped launch the careers of several successful designers including Mike Whittaker (designer of the MW series microlights) and Philip Lambert (designer of the Lambert Mission light aeroplane)
What are the prizes?	Usually support in developing ideas further after the competition. The specific theme usually changes between competitions
Where's the website?	https://www.aerosociety.com/get-involved/specialist-groups/business-general-aviation/general-aviation/

What's the competition?	Shell Eco-Marathon
Who organizes it?	Shell Global
When and where?	Asia: March USA: April Europe: May–July Brazil: November Various prior entry deadlines and testing events occur in the months before the final competitions
Who can enter?	Teams of young engineers
What's it all about?	Design, build and test an ultra energy efficient car, to deliver the longest possible driving distance on one litre of fuel
What are the prizes?	Multiple cash and trophy prizes in categories including energy efficiency, design, communications, innovation and safety
Where's the website?	https://www.shell.com/energy-and-innovation/shell-ecomarathon.html

What's the competition?	Spaceport America Cup, including The Intercollegiate Rocket Engineering Competition, and the Space Dynamics Laboratory Payload Challenge
Who organizes it?	Experimental Sounding Rocket Association
When and where?	Annual, New Mexico, USA
Who can enter?	Anybody, but mainly aimed at studentteams from around the world
What's it all about?	Space technology-related practical projects
What are the prizes?	Up to $750
Where's the website?	http://www.soundingrocket.org/what-is-irec.html

What's the competition?	Starpack
Who organizes it?	Institute of Materials, Minerals and Mining and The Packaging Society
When and where?	Global and annual
Who can enter?	Undergraduate students
What's it all about?	Developing and presenting innovative solutions in packaging
What are the prizes?	Trophies and publicity
Where's the website?	http://www.iom3.org/starpack/student-starpack-awards-0

What's the competition?	Telegraph STEM awards
Who organizes it?	Daily Telegraph, Babcock
When and where?	United Kingdom. Entries October–February Shortlisting and initial presentations early March Category winner presentations late March Winners announced June
Who can enter?	Any undergraduate student of any nationality studying at a UK university. Either as individuals or teams
What's it all about?	Solving one of several problems presented by industry sponsors
What are the prizes?	Overall winner: work experience and mentoring programme and £25,000 Category winners: work experience programmes
Where's the website?	http://www.telegraph.co.uk/education/stem-awards/

What's the competition?	Toyota Forklifts Handling Challenge
Who organizes it?	Toyota Europe
When and where?	Submitted and judged online
Who can enter?	Undergraduate or Graduatestudents. The competition is mainly aimed at design related subjects
What's it all about?	An annual competition to solve a technological goods handling problem
What are the prizes?	Up to €5,000
Where's the website?	https://tldc.toyota-forklifts.eu/pages/the-challenge

Further Resources

The sources below may well have further information on competitions available for entry.

- Shell (the energy company) often organize and sponsor engineering design competitions under the "Ideas 360" label; which may appear in various places online.
- Your engineering institution, at international, national and branch level (and if you haven't joined one, do so, this is really important to your development and career).

Fig. C.1 Some of our students from Brunel University London with their Formula Student racecar

– (US) National Aeronautics and Space Administration (NASA) website: https://
  www.nasa.gov/audience/forstudents/stu-competitions-current-opps.html
– Design on demand website: https://desall.com/Contests
– https://studentcompetitions.com/

Fig. C.2 Student engineer competitors at the international submarine competition (courtesy of United States Navy, NSWC Carderock Division)

– Born to Engineer (UK) competitions website: https://www.borntoengineer.com/resources/engineering-competitions
– (USA) National Academy of Engineering, Engineer Girl Website Competitions Page: https://www.engineergirl.org/250/Contests

C.3 Presentation Competitions

What's the competition?	Aerospace Final Year ProjectCompetition
Who organizes it?	(UK) Association of Aerospace Universities
When and where?	Main competition will be on a Friday in late October at a UK University, for projects from the previous academic year
Who can enter?	Any AAU member university can enter up to two projects per year

(continued)

(continued)

What's it all about?	Universities nominate their best aerospace engineering studentprojects at BEng, MEng or MSc level. Students make 20 min presentations before judges
What are the prizes?	First prize: £500 Two runners up prizes: £300 John Farley trophy for the project making the greatest contribution to aviation safety
Where's the website?	http://www.aau.ac.uk/awards.htm

What's the competition?	Engineering New Zealand Annual PresentationCompetition
Who organizes it?	Engineering New Zealand
When and where?	Annual competition in
Who can enter?	New Zealand based engineering students and young engineers
What's it all about?	Giving presentations on selected topics
What are the prizes?	Small
Where's the website?	https://www.engineeringnz.org/courses-events/event/young-engineers-annual-presentation-competition/

What's the competition?	IOM3 Young Persons' Lecture Competition
Who organizes it?	Institute of Materials, Minerals and Mining and the Worshipful Company of Armourers and Braziers
When and where?	London, April
Who can enter?	Persons aged under 28 on 1st June
What's it all about?	Competitors give a lecture of 15 min on a topic related to materials, minerals, mining, packaging, clay technology and wood science and engineering
What are the prizes?	Not known
Where's the website?	http://www.iom3.org/young-persons-lecture-competition

What's the competition?	Present Around The World
Who organizes it?	Institution of Engineering and Technology (IET)
When and where?	Globally: preliminary/local network rounds: September–May National area competition: April–June Regional competition: July–August Global final: November (London, UK)
Who can enter?	Young professionals and students within engineering; 18–30 years of age
What's it all about?	10-minute presentation. Judging is based on presentation skills and technical content

(continued)

(continued)

What are the prizes?	Global final winner: £1,000; runner up £500 Regional winner: £400; runner up £300 National winner: £300; runner up £200 Local network winner: £150; runner up £100
Where's the website?	www.theiet.org/patw

What's the competition?	Source America Design Challenge
Who organizes it?	Source America
When and where?	Project submission: January Semi-finals: February Final events: Spring
Who can enter?	Teams of high school and college students working with a non-profit agency and at a disabledperson (termed the "subject matter expert" or SME)
What's it all about?	Pairing students with technical skills up with a disabledperson to jointly develop a process or invention that will improve disabled people's ability to thrive in the workplace
What are the prizes?	Expenses are paid for up to 5 students and a coach to attend competitions. Mentoring is also available for teams Multiple prizes up to $8,000 split between the team, $6,000 for the college and $3,000 for the collaborating non-profit agency
Where's the website?	https://www.sourceamerica.org/design-challenge

What's the competition?	United Arab Emirates, Royal Aeronautical Society Young Persons Lecture Competition
Who organizes it?	Royal Aeronautical Society
When and where?	Globally, usually organized by local society branches
Who can enter?	Aeronautical Engineers
What's it all about?	Aeronautical Engineers aged under 30, and under 22 give specialist lectures to a jury
What are the prizes?	Varies, usually small
Where's the website?	www.aerosociety.com but check with your local branch

Appendix D
About Mr William Heath Robinson

Most of the illustrations in this book were drawn by Mr William Heath Robinson, who until now is probably best known to most modern engineering students as the illustrator of the classic children's book "The Incredible Adventures of Professor Branestawm".

William Heath Robinson was born on 31 March 1872 in London into a family of artists and craftsmen. His grandfather, Thomas Robinson, had known and bound books for George Stephenson, the inventor of the locomotive steam engine, before becoming a successful engraver, and his father—also called Thomas Robinson—made his living illustrating a weekly newspaper in the days before the technology existed for newspapers to carry photographs. William records in his autobiography [1] that he and his brothers and sisters attended unremarkable local schools, but spent much of their time at home drawing, usually with chalk on slate, when not playing games such as conkers (Fig. D.1) and marbles in the streets.

At the age of 15, encouraged by his family, William left school, as his older brother Tom already had, to attend art school in the nearby London suburb of Islington. This successfully prepared him, on his second attempt, to gain entry to the Royal Academy Schools—where he studied regularly at the British Museum, drawing the artefacts there. As his skill developed, he tried to earn his living as a landscape painter—painting particularly Hampstead Heath, a very popular nineteenth-century artistic location. Commercially, this endeavour failed him, and he only ever sold a single one of these paintings—to a friend of his mother's.

Now in his early 20s, and needing to earn a living, William followed his brothers into commercial art. Working from his father's studio in a tiny room with no windows but a skylight, he started to produce book and magazine illustrations. He proved to have considerable talent for this with straightforward workmanlike illustrations, often to children's stories. In 1900, aged 24, he received his best commission to date—illustrating a book of poems by Edgar Allan Poe (Fig. D.2) [2], along with the book that probably transformed his career—a children's story about an inventor which he also wrote called *The Adventures of Uncle Lubin* (still in print at reference [3]). Uncle Lubin made him enough money to finally marry his fiancée, Miss Josephine Latey, in 1903, with whom he had their first of five children, Joan, a year later.

© Springer Nature Switzerland AG 2020
P. Gratton and G. Gratton, *Achieving Success with the Engineering Dissertation*,
https://doi.org/10.1007/978-3-030-33192-4

Fig. D.1 Playing conkers, still a popular childhood game in Britain. (by Mr William Heath Robinson)

Fig. D.2 Illustration for the poem "The Raven" by Edgar Allan Poe (by Mr William Heath Robinson)

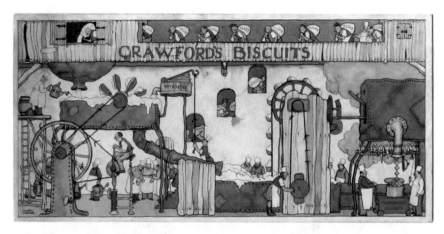

Fig. D.3 An advertisement for Crawford's Biscuits, showing Heath Robinson's developing and popular humorous style (by Mr William Heath Robinson)

"Uncle Lubin" came to the attention of many people—HG Wells was a fan of it, but in particular it brought invitations to draw advertisements, initially for the *Lamson Paragon Supply Co. Ltd* with whom he went on to enjoy an eight-year relationship. These advertising drawings built upon *Uncle Lubin* contained a great deal of humour (Fig. D.3), and led eventually to fanciful and complex inventions that started to enter his artwork from about 1915 and regularly caught the public's imagination—they also led to commissions for similar work in magazines such as *The Strand* (best known as where Conan Doyle's Sherlock Holmes stories were first published) and the new tabloid newspaper *The Daily Sketch*.

Heath Robinson continued into the 1910s to draw both serious work, including illustrations for editions of *don Quixote*, *The Arabian Nights* and various works by William Shakespeare—and his humorous illustrations full of complex and usually impossible inventions. However, he found that his name had become a byword for the latter, making it increasingly difficult to sell his more serious art, although he never ceased to do so.

During the First World War, which Britain entered in 1914, Heath Robinson—now in his 40s—was considered too old for war service, so found a new theme, which was to portray the war in a humorous manner (Fig. D.4) [4]. At a time when there was little humour to be found, this was very well received. He received, and usually replied to, many letters from servicemen in all theatres of war often sharing their experiences and suggesting he might use them in his artwork; on at least one occasion government scientists wrote to him with a photograph of a piece of test equipment (a device for calibrating petrol jets, presumably for carburettors) which they had named after him. Towards the end of the war, he was commissioned by an American syndicate also to prepare a series of humorous drawings of the American army in France, and spent some time there around the fronts: at one point being arrested as a possible spy due to his sketching without the right permit. Immediately

Fig. D.4 Heath Robinson humourized the very serious U-boat and Zeppelin threats with his illustration of the fictional "Subzeppmarinellin". In [1] he recounts experiencing both Zeppelin and aeroplane raids himself, as well as being on board a troop ship being hunted by submarines. (by Mr William Heath Robinson)

after the end of the war, he found a very similar artistic avenue as he drew weapons of war being re-purposed for peace, such as re-using old tanks as buses.

Subsequently, William Heath Robinson continued to pursue both his serious and humorous commercial artwork, and even gave a few BBC broadcasts telling listeners how to produce their own art. He found particular interests in producing drawings of aspects of British industry (Fig. D.5) such as structural engineering and coal mining. He wrote in his autobiography [1] "Steel girders, Swiss rolls, welding, toffee, paper making, marmalade, asbestos cement, beef essence, motor spirit and lager beer were among the many and diverse subjects to be treated. As the principles of mechanics are always the same, this variety did not matter very much." By the 1930s, he was a household name in much of the English speaking world and received many invitations for commissions, including illustrating in 1933 a book that many engineers will be familiar with from their childhood: *The Incredible Adventures of Professor* Branestawm [5], although later books about the famous absent-minded professor were illustrated by other artists.

During World War II, Heath Robinson (Fig. D.6), now approaching 70 but still not retired, repeated his First World War role as a war cartoonist—producing many inventive ideas concerning methods of warfare, new military gadgets, invasion prevention camouflage and satirical responses to German propaganda, rationing and many other topics of the day. This, his final artistic phase, was extremely productive [6]. One of the early decryption machines at the British government codebreaking centre at Bletchley Park was also named after Heath Robinson, although given the secrecy of the establishment, it is unlikely that he ever knew this.

William Heath Robinson died on 13 September 1944, aged 72, leaving his name as a phrase for unlikely and over-complicated inventions (terms like "it's a bit Heath Robinson" are still well known) and many thousands of items of artwork to posterity.

If you are near London, you can learn more about Heath Robinson's life and career, and see much of his original artwork, and sometimes several sculptures based upon his art, by visiting the Heath Robinson Museum [7] in Pinner (very close to Pinner underground station, on the Metropolitan Line, and of course to former Heath Robinsonfamily homes); the museum is run jointly by the Heath Robinson Trust and the local community of Pinner. A nearby pizza restaurant is also decorated with original sculptures based upon his work. Many books either of Heath Robinson cartoons or illustrated with his artwork are also still widely available—including from the Museum's shop in Pinner.

THE CHANGE OVER FROM BROAD TO NARROW GAUGE

Fig. D.5 One of a series of illustrations produced for the Great Western Railway Company

Fig. D.6 Mr William Heath
Robinson, Self Portrait

References

[1] Heath Robinson. W., My Line of Life, Blackie and Son, London, 1938.
[2] Poe. E. A., The poems of Edgar Allan Poe: illustrated by W Heath Robinson, George Bell and
    Sons (London) 1900
[3] Heath Robinson. W., The Adventures of Uncle Lubin, Dover, London, 2013
[4] Heath Robinson. W., Heath Robinson's Great War: The Satirical Cartoons 2nd edn., Bodleian
    Library, 2015
[5] Hunter. N., The Incredible Adventures of Professor Branestawm, Vintage, 2013
[6] Heath Robinson. W., Heath Robinson's Second World War: The Satirical Cartoons, Bodleian
    Library 2016
[7] https://www.heathrobinsonmuseum.org/

Index

© Springer Nature Switzerland AG 2020
P. Gratton and G. Gratton, *Achieving Success with the Engineering Dissertation*,
https://doi.org/10.1007/978-3-030-33192-4

273

Printed in the United States
By Bookmasters